SOLUTIONS MANUAL

for the

MECHANICAL ENGINEERING REFERENCE MANUAL

Ninth Edition

Michael R. Lindeburg, P.E.

PROFESSIONAL PUBLICATIONS, INC.
Belmont, CA 94002

In the ENGINEERING LICENSING EXAM AND REFERENCE SERIES

Engineer-In-Training Reference Manual
  EIT Review Manual
  Engineering Fundamentals Quick Reference Cards
  Engineer-In-Training Sample Examinations
  Mini-Exams for the E-I-T Exam
  1001 Solved Engineering Fundamentals Problems
  Fundamentals of Engineering Exam Study Guide
  Diagnostic F.E. Exam for the Macintosh
  Fundamentals of Engineering Video Series: Thermodynamics
Civil Engineering Reference Manual
  Civil Engineering Quick Reference Cards
  Civil Engineering Sample Examination
  Civil Engineering Review Course on Cassettes
  101 Solved Civil Engineering Problems
  Seismic Design of Building Structures
  Seismic Design Fast
  345 Solved Seismic Design Problems
  Timber Design for the Civil P.E. Exam
  Fundamentals of Reinforced Masonry Design
  246 Solved Structural Engineering Problems
Mechanical Engineering Reference Manual
  Mechanical Engineering Quick Reference Cards
  Mechanical Engineering Sample Examination
  101 Solved Mechanical Engineering Problems
  Mechanical Engineering Review Course on Cassettes
  Consolidated Gas Dynamics Tables
  Fire and Explosion Protection Systems
Electrical Engineering Reference Manual
  Electrical Engineering Quick Reference Cards
  Electrical Engineering Sample Examination
Chemical Engineering Reference Manual
  Chemical Engineering Quick Reference Cards
  Chemical Engineering Practice Exam Set
Land Surveyor Reference Manual
  Land Surveyor-In-Training Sample Examination
  1001 Solved Surveying Fundamentals Problems

In the GENERAL ENGINEERING and CAREER ADVANCEMENT SERIES

How to Become a Professional Engineer
Getting Started as a Consulting Engineer
The Expert Witness Handbook: A Guide for Engineers
Engineering Your Job Search
Engineering Your Start-Up
Intellectual Property Protection: A Guide for Engineers
High-Technology Degree Alternatives
Metric in Minutes
Engineering Economic Analysis
Engineering Law, Design Liability, and Professional Ethics
Engineering Unit Conversions

SOLUTIONS MANUAL for the
MECHANICAL ENGINEERING REFERENCE MANUAL
Ninth Edition

Printed in the United States of America

ISBN: 0-912045-84-1

Professional Publications, Inc.
1250 Fifth Avenue, Belmont, CA 94002
(415) 593-9119

Current printing of this edition: 3

Revised and reprinted in 1995.

TABLE OF CONTENTS

Notice to Examinees

Do not copy, memorize, or distribute problems from the Principles and Practice of Engineering (P&P) Examination. These acts are considered to be exam subversion.

The P&P examination is copyrighted by the National Council of Examiners for Engineering and Surveying. Copying and reproducing P&P exam problems for commercial purposes is a violation of federal copyright law. Reporting examination problems to other examinees invalidates the examination process and threatens the health and welfare of the public.

PREFACE AND ACKNOWLEDGMENTS TO THE NINTH EDITION

The *Mechanical Engineering Reference Manual* was the first PE examination book I ever wrote. Accordingly, its accompanying solutions manual was the first PE solutions manual I wrote. In addition to being laboriously handwritten in my nearly-illegible scrawl, the solutions manual suffered from personal idiosyncrasies and nearly almost every conceivable inconsistency in presentation and style. It is hard to believe that the handwritten solutions manual survived ten years and nine editions.

Well, that has now changed. This new 9th edition has been edited for consistency in style, typeset, and professionally illustrated. I have used this major revision as a chance to include many alternate solutions to problems that can be solved in different ways. I won't enumerate all the other specific changes and improvements. They're only of historical interest, and most engineers are not historians. Taken all together, though, the changes will make the *Solutions Manual to the Mechanical Engineering Reference Manual* much, much easier for you to use.

In producing these solutions, I have introduced a few inconsistencies with the existing *Mechanical Engineering Reference Manual*. Specifically, scientific notation is written in normal fashion, unlike the "EE" format used in the *Reference Manual*. Specific weight (γ in lbf/ft^3) is distinguished from density (ρ in lbm/ft^3). Also, dimensionless numbers are identified by their first two letters (e.g., Re for Reynolds number), unlike in the *Reference Manual* which subscripts the first two letters (e.g., N_{Re}). I doubt that these inconsistencies will be of any concern to you.

As usual, I am indebted to the production staff at Professional Publications for their assistance in modernizing this book. What impressed me the most about this book were the on-the-fly improvements made at all stages of the production process. The depth of engineering knowledge exhibited by this non-engineering staff is phenomenal. Sylvia M. Osias and Mary Christensson did the typesetting. The illustrations were produced by Gregory J. McGreevy and imported directly into the manuscript. Jessica R. Whitney did the proofreading. The schedule was produced and the work overseen by Jessica R. Whitney. Thanks to you all.

As always, your suggestions for improvement are welcome. That's the purpose of the response card in the back of this book. You will never know how many significant improvements, shortcuts, and insights were contributed by readers like yourself, but you can contribute to someone else's learning if you wish.

Michael R. Lindeburg, PE
Belmont, CA
November 1994

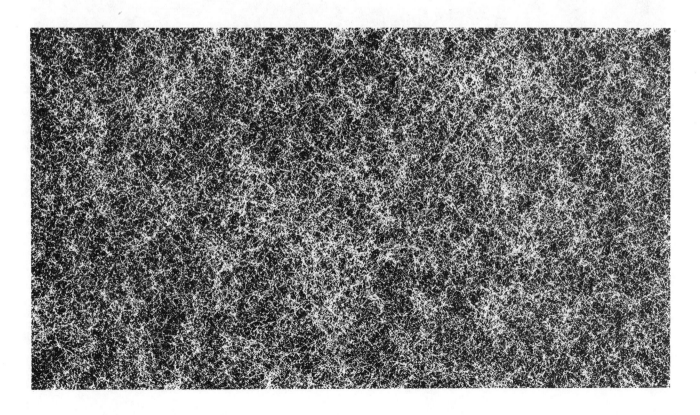

MATHEMATICS

Warmups

1. $\displaystyle\sum_{j=1}^{5}(j+1)^2 - 1 = (1+1)^2 - 1 + (2+1)^2 - 1$
$$+ (3+1)^2 - 1 + (4+1)^2 - 1$$
$$+ (5+1)^2 - 1$$
$$= 2^2 + 3^2 + 4^2 + 5^2 + 6^2 - 5$$
$$= \boxed{85}$$

2. The actual value is
$$y(2.7) = (3)(2.7)^{0.93} + 4.2 = 11.756$$
$$y(2) = (3)(2)^{0.93} + 4.2 = 9.916$$
$$y(3) = (3)(3)^{0.93} + 4.2 = 12.534$$

The estimated value is
$$9.916 + (0.7)(12.534 - 9.916) = 11.749$$

The error is
$$\frac{11.756 - 11.749}{11.756} = \boxed{0.0006 \ (0.06\%)}$$

3. Let d be the diameter.
$$V_{\text{sphere}} = \left(\frac{4}{3}\right)\pi r^3 = \left(\frac{4}{3}\right)\pi\left(\frac{d}{2}\right)^3 = 0.524d^3$$
$$V_{\text{cone}} = \left(\frac{\pi}{3}\right)r^2 h = \left(\frac{\pi}{3}\right)\left(\frac{d}{2}\right)^2 h = 0.262d^2 h$$
$$0.524d^3 = 0.262d^2 h$$
$$h = \boxed{2.00d}$$

4.

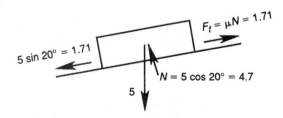

5. Expand by the second column.
$$-2\begin{vmatrix} 4 & 3 \\ 9 & 5 \end{vmatrix} = (-2)(20-27) = \boxed{14}$$

6. $\qquad \left(\dfrac{5}{9}\right)(250°\text{F} - 32) = \boxed{121.1°\text{C}}$
$$250°\text{F} + 460 = \boxed{710°\text{R}}$$

7. $K = 1.71 \times 10^{-9} \ \dfrac{\text{BTU}}{\text{ft}^2\text{-hr-}°\text{R}^4}$

$$\times \frac{\left(1.71 \times 10^{-9} \ \dfrac{\text{BTU}}{\text{ft}^2\text{-hr-}°\text{R}^4}\right)\left(17.57 \ \dfrac{\text{W-min}}{\text{BTU}}\right)\times\left(\dfrac{1}{60}\dfrac{\text{hr}}{\text{min}}\right)}{\left(0.3048 \ \dfrac{\text{m}}{\text{ft}}\right)^2\left(\dfrac{5}{9}\dfrac{\text{K}}{°\text{R}}\right)^4}$$

$$= \boxed{5.66 \times 10^{-8} \ \text{W/m}^2\cdot\text{K}^4}$$

8. $y = 6 + (0.75)(2-6) = \boxed{3.0}$

9. The slope is
$$\frac{9.5 - 3.4}{8.3 - 1.7} = 0.924$$

Using the first point,
$$\boxed{y - 3.4 = (0.924)(x - 1.7)}$$

10. Let x be the number of elapsed periods of 0.1 sec, and let y_x be the amount present after x periods.

$$y_1 = 1.001y_0$$
$$y_2 = (1.001)^2 y_0$$
$$y_n = (1.001)^n y_0$$
$$\frac{y_x}{y_0} = 2 = (1.001)^n$$
$$\log(2) = n\log(1.001)$$
$$n = 693.5 \text{ periods}$$
$$t = \boxed{69.35 \text{ sec}}$$

Concentrates

1. Rearrange.

$$
\begin{aligned}
x + y \quad\ \ &= -4 \\
x \quad\ + z &= \ \ 1 \\
3x - y + 2z &= \ \ 4
\end{aligned}
$$

Use Cramer's rule.

$$
\begin{vmatrix} 1 & 1 & 0 \\ 1 & 0 & 1 \\ 3 & -1 & 2 \end{vmatrix} = (1)\begin{vmatrix} 0 & 1 \\ -1 & 2 \end{vmatrix} - (1)\begin{vmatrix} 1 & 1 \\ 3 & 2 \end{vmatrix}
$$

$$
= (0+1) - (2-3)
$$

$$
= 1 + 1 = 2
$$

$$
\begin{vmatrix} -4 & 1 & 0 \\ 1 & 0 & 1 \\ 4 & -1 & 2 \end{vmatrix} = -2
$$

$$
x^* = \frac{-2}{2} = \boxed{-1}
$$

$$
\begin{vmatrix} 1 & -4 & 0 \\ 1 & 1 & 1 \\ 3 & 4 & 2 \end{vmatrix} = -6
$$

$$
y^* = \frac{-6}{2} = \boxed{-3}
$$

$$
\begin{vmatrix} 1 & 1 & -4 \\ 1 & 0 & 1 \\ 3 & -1 & 4 \end{vmatrix} = 4
$$

$$
z^* = \frac{4}{2} = \boxed{2}
$$

2. Plot the data points to determine if a linear relationship exists between the variables.

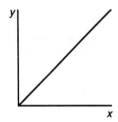

Use linear regression.

$$ n = 7 $$

$$ \sum y = 12{,}550 \qquad\qquad \sum_y = 12{,}300 $$

$$ \overline{x} = 1792.9 \qquad\qquad \overline{y} = 1757.1 $$

$$ \sum x^2 = 3.117 \times 10^7 \qquad \sum y^2 = 3.017 \times 10^7 $$

$$ \left(\sum x\right)^2 = 1.575 \times 10^8 \qquad \left(\sum y\right)^2 = 1.513 \times 10^8 $$

$$ \sum xy = 3.067 \times 10^7 $$

$$
m = \frac{(7)(3.067 \times 10^7) - (12{,}550)(12{,}300)}{(7)(3.117 \times 10^7) - (12{,}550)^2} = 0.994
$$

$$
b = 1757.1 - (0.994)(1792.9) = -25.0
$$

So,

$$
y = 0.994x - 25.0
$$

The correlation coefficient is

$$
r = \frac{(7)(3.067 \times 10^7) - (12{,}500)(12{,}300)}{\sqrt{\begin{aligned}&[(7)(3.117 \times 10^7) - (12{,}500)^2] \\ &\times [(7)(3.017 \times 10^7) - (12{,}300)^2]\end{aligned}}}
$$

$$
\approx \boxed{1.00}
$$

3. Plotting the data shows that the relationship between the variables is nonlinear.

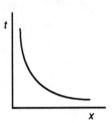

This appears to be an exponential with the form

$$
t = be^{ms}
$$

$$
\log t = b + ms
$$

Try making the variable transformation $R = \log t$.

s	R
20	1.633
18	2.149
16	2.585
14	3.041

$$ n = 4 $$

$$ \sum s = 68 \qquad\qquad \sum R = 9.408 $$

$$ \overline{s} = 17 \qquad\qquad \overline{R} = 2.352 $$

$$ \sum s^2 = 1176 \qquad\qquad \sum R^2 = 23.215 $$

$$ \left(\sum s\right)^2 = 4624 \qquad \left(\sum R\right)^2 = 88.51 $$

$$ \sum sR = 155.28 $$

$$
m = \frac{(4)(155.28) - (68)(9.408)}{(4)(1176) - (68)^2} = -0.2328
$$

$$
b = 2.352 - (-0.2328)(17) = 6.3096
$$

So,

$$
R = 6.3096 - 0.2328s
$$

$$
\log t = 6.3096 - 0.2328s
$$

PROFESSIONAL PUBLICATIONS, INC. ● Belmont, CA

4. This is a first-order linear differential equation.

$$u = \exp\left(\int -1\,dx\right) = e^{-x}$$
$$y = e^x \left(2 \int e^{-x} x e^{2x}\,dx + C\right)$$
$$= e^x \left(2xe^x - 2e^x + C\right)$$

But $y = 1$ when $x = 0$.

$$1 = (1)[0 - 2 + C] \quad \text{or} \quad C = 3$$

$$y = \boxed{2e^{2x}(x-1) + 3e^x}$$

5. Solve the characteristic quadratic equation.

$$R^2 - 4R - 12 = 0$$
$$R = 6, -2$$

$$y = \boxed{a_1 e^{6x} + a_2 e^{-2x}}$$

6. $\quad x_t$ = mass of salt in tank at time t, lbm

$\quad x_0 = 60$ lbm

$\quad x'$ = rate at which salt content changes

$\quad 2$ = lbm salt entering each minute

$\quad 3$ = gal leaving each minute

The amount of salt leaving each minute is

$$(3)\binom{\text{concentration in}}{\text{lbm/gal}} = (3)\left(\frac{\text{salt content}}{\text{volume}}\right)$$
$$= (3)\left(\frac{x}{100 - t}\right)$$
$$x' = 2 - (3)\left(\frac{x}{100 - t}\right)$$
$$x' + \frac{3x}{100 - t} = 2$$

This is a first-order linear differential equation.

$$u = \exp\left(3\int \frac{dt}{100 - t}\right) = (100 - t)^{-3}$$
$$x = (100 - t)^3 \left[2 \int \frac{dt}{(100 - t)^3} + k\right]$$
$$= 100 - t + k(100 - t)^3$$

$x = 60$ at $t = 0$, so $k = -0.00004$.

$$x = 100 - t - (0.00004)(100 - t)^3$$
$$x_{60} = 100 - 60 - (0.00004)(100 - 60)^3$$
$$= \boxed{37.44 \text{ lbm}}$$

7. If c is positive, then $n(\infty) = \infty$, which is contrary to the data given. Therefore, $c \le 0$.

If $c = 0$, then $n(\infty) = a/(1+b) = 100$, which is possible, depending on a, b.

If $c = 0$, then $n(0) = a/(1 + b) = 10$, which conflicts with the previous step.

So $c < 0$, then $n(\infty) = a$, so $a = \boxed{100}$

$$n(0) = \frac{a}{1 + b} = \frac{100}{1 + b} = 10$$

$$b = \boxed{9}$$

$$\frac{dn}{dt} = (-100)(1 + 9e^{ct})^{-2}(9)e^{ct}(c)$$

If $t = 0$, then $c = \boxed{-0.0556}$

8. $\quad \dfrac{dy}{dx} = 3x^2 - 18x$

$\quad 3x^2 - 18x = 0 \quad$ [at all extreme points]

$\quad x^2 - 6x = 0$

This is satisfied at $x = 0$, $x = 6$.

$$\frac{d^2y}{dx^2} = 6x - 18$$
$$6x - 18 = 0 \quad \text{[at inflection points]}$$
$$x = 3$$

$$\boxed{x = 3 \text{ is an inflection point}}$$

$$(6)(0) - 18 = -18$$

$$\boxed{x = 0 \text{ is a maximum}}$$

$$(6)(6) - 18 = 18$$

$$\boxed{x = 6 \text{ is a minimum}}$$

9. The energy contained in one gram of any substance is

$$E = mc^2 = (0.001 \text{ kg})\left(3 \times 10^8 \, \frac{\text{m}}{\text{s}}\right)^2$$
$$= 9 \times 10^{13} \text{ J}$$

$$(9 \times 10^{13} \text{ J}) \left(\frac{1 \text{ kJ}}{1000 \text{ J}} \right)$$

$$\times \left(0.9478 \, \frac{\text{BTU}}{\text{kJ}} \right) = 8.53 \times 10^{10} \text{ BTU}$$

$$\text{tons} = \frac{8.53 \times 10^{10} \text{ BTU}}{\left(13{,}000 \, \dfrac{\text{BTU}}{\text{lbm}} \right) \left(2000 \, \dfrac{\text{lbm}}{\text{ton}} \right)}$$

$$= 3281 \text{ tons}$$

10.

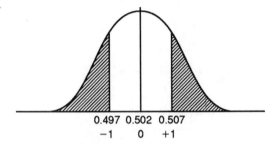

$$\begin{array}{ccc} 0.497 & 0.502 & 0.507 \\ -1 & 0 & +1 \end{array}$$

The standard normal variables are

$$z_1 = \frac{0.502 \text{ in} - 0.497 \text{ in}}{0.005 \text{ in}} = 1$$

$$z_2 = \frac{0.507 \text{ in} - 0.502 \text{ in}}{0.005 \text{ in}} = 1$$

(a) $p\{\text{defective}\} = (2)(0.5 - 0.3413) = \boxed{0.3174}$

(b) $p\{3, 2\} = \left[\dfrac{3!}{(3-2)! \, 2!} \right] (0.3174)^2 (1 - 0.3174)$

$$= \boxed{0.2063}$$

(c) $(8)(200)(0.3174) = \boxed{507.8}$

11. The range of speeds is 48 mph − 20 mph = 28 mph. Since there are not a lot of observations, 10 cells would be best. Choose the cell width as 28 mph/10 ≈ 3 mph.

(a) and (d)

interval	midpoint	freq.	cum. freq.	cum. fraction
20–22	21	1	1	0.03
23–25	24	3	4	0.10
26–28	27	5	9	0.23
29–31	30	8	17	0.43
32–34	33	3	20	0.50
35–37	36	4	24	0.60
38–40	39	3	27	0.68
41–43	42	8	35	0.88
44–46	45	3	38	0.95
47–49	48	2	40	1.00

(b) and (c)

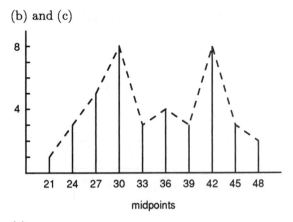

(e)

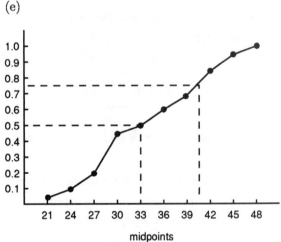

(f) Use the cumulative distribution graph. For 75% (0.75), the cell midpoint is approximately 40 mph.

(g) Use the cumulative distribution graph to find the midpoint for 50%. This occurs at approximately 33 mph. The distribution is bimodal (i.e., has two peaks). The two modes are 30 mph and 42 mph.

$$\sum x_i = 1390$$

$$\overline{x} = \frac{1390 \text{ mph}}{40} = \boxed{34.75 \text{ mph}}$$

(h) $$\sum x^2 = 50{,}496$$

$$\sigma = \sqrt{\frac{50{,}496}{40} - \left(\frac{1390}{40} \right)^2}$$

$$= \boxed{7.405 \text{ mph}}$$

(i) $s = \sigma \sqrt{\dfrac{n}{n-1}} = 7.405 \sqrt{\dfrac{40}{39}} = \boxed{7.500 \text{ mph}}$

(j) $s^2 = \boxed{56.25 \text{ mph}^2}$

(f) $\boxed{\text{Float is the same as slack} = 0.}$

12. No contract deadline was given, so assume 36 as a scheduled time. Look for a path where $LS - ES = 0$ everywhere.

(a) and (b)

The critical path is shown by the darker line.

13. To solve this as a regular CPM problem, it is necessary to calculate t_{mean} and σ for each activity. For activity A,

$$t_{mean} = \left(\tfrac{1}{6}\right)\left[1 + (4)(2) + 5\right] = 2.33$$
$$\sigma_A = \left(\tfrac{1}{6}\right)(5 - 1) = 0.67$$

The following table is generated in the same manner.

activity	t_{mean}	σ
start	0	0
A	2.33	0.67
B	10.5	2.17
C	11.83	2.17
D	4.17	0.83
finish	0	0
	$\overline{28.83}$	

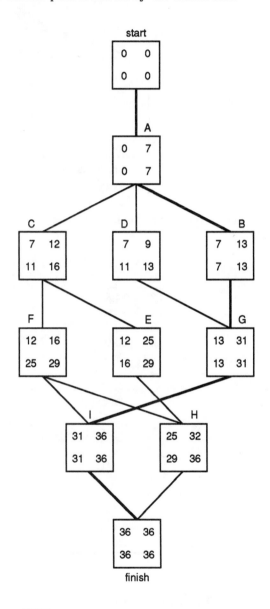

(c) $\boxed{36}$

(d) $\boxed{36}$

(e) $\boxed{0}$

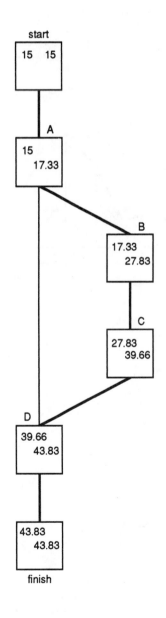

By observation, the critical path is start-A-B-C-D-finish. The project variance is

$$\sigma^2 = (0.67)^2 + (2.17)^2 + (2.17)^2 + (0.83)^2 = 10.56$$

The project standard deviation is

$$\sigma = \sqrt{10.56} = 3.25$$

Assume the completion times are normally distributed with a mean of 28.83 and a standard deviation of 3.25. The standard normal variable is

$$z = \frac{28.83 + 15 - 42}{3.25} = 0.56$$

The area under the tail of the normal curve for $z = 0.56$ is 0.2123, so

$$0.5 - 0.2123 = \boxed{0.2877 \quad (28.77\%)}$$

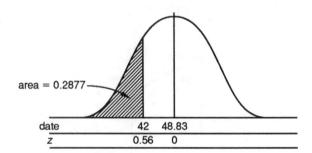

14. $\lambda = 20$ cars.

(a) $p\{x = 17\} = f(17) = \dfrac{e^{-20}(20)^{17}}{17!} = 0.076$

(b) $p\{x \le 3\} = f(0) + f(1) + f(2) + f(3)$

$$= \frac{e^{-20}(20)^0}{0!} + \frac{e^{-20}(20)^1}{1!}$$

$$+ \frac{e^{-20}(20)^2}{2!} + \frac{e^{-20}(20)^3}{3!}$$

$$= (2.06 \times 10^{-9}) + (4.12 \times 10^{-8})$$
$$+ (4.12 \times 10^{-7}) + (2.75 \times 10^{-6})$$

$$= \boxed{3.2 \times 10^{-6}}$$

15.

$$\mu = \frac{1}{23}$$

$$p\{x > 25\} = 1 - F(25) = e^{-\left(\frac{1}{23}\right)(25)}$$

$$= \boxed{0.337}$$

16. Plot the data to see if they are linear.

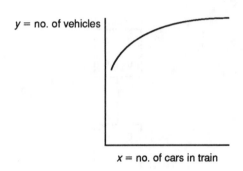

This is not linear. Try the form $y = a + bz$ where $z = \log_{10} x$.

z	y
0.3	14.8
0.7	18.0
0.9	20.4
1.08	23.0
1.43	29.9

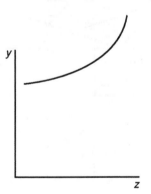

This is not linear either. Try the form $y = a + bw$ where $w = z^2 = (\log_{10} x)^2$.

w	y
0.09	14.8
0.49	18.0
0.81	20.4
1.17	23.0
2.04	29.9

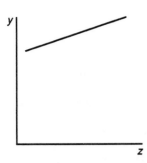

This looks linear.

$$\sum w_i = 4.6 \qquad\qquad \sum y_i = 106.1$$
$$\left(\sum w_i\right)^2 = 21.16 \qquad \left(\sum y_i\right)^2 = 11{,}257.2$$
$$\sum w_i^2 = 6.43 \qquad\qquad \sum y_i^2 = 2382.2$$
$$\overline{w} = \frac{4.6}{5} = 0.92 \qquad \overline{y} = \frac{106.1}{5} = 21.22$$

$$\sum w_i y_i = 114.58$$

$$m = \frac{(5)(114.58) - (4.6)(106.1)}{(5)(6.43) - 21.16}$$
$$= 7.72$$
$$b = 21.22 - (7.72)(0.92) = 14.12$$

So,
$$y = 14.12 + 7.72w$$
$$= 14.12 + 7.72z^2$$
$$= 14.12 + (7.72)(\log_{10} x)^2$$

The correlation coefficient is

$$r = \frac{(5)(114.58) - (4.6)(106.1)}{\sqrt{[(5)(6.43) - 21.16]\,[(5)(2382.2) - 11{,}257.2]}}$$
$$\approx 0.999$$

(The transformation $y = a + b\sqrt{x}$ yields $y = 9.11 + 4\sqrt{x}$ and $r = 0.999$, equally good.)

17. (a) Use the characteristic equation method to solve the homogeneous case. (It is much quicker to use Laplace transforms, however.)

$$x'' + 2x' + 2x = 0 \quad \text{[differential equation]}$$
$$R^2 + 2R + 2 = 0 \quad \text{[characteristic equation]}$$

Complete the square to find R.

$$R^2 + 2R = -2$$
$$(R+1)^2 = -2 + 1$$
$$R + 1 = \pm\sqrt{-1}$$
$$R = -1 \pm i$$

So,
$$x(t) = A_1 e^{-t}\cos t + A_2 e^{-t}\sin t$$

Use the initial conditions to find A_1 and A_2.

$$x(0) = 0$$
$$0 = A_1(1)(1) + A_2(1)(0)$$
$$A_1 = 0$$

Differentiating the solution,

$$x'(t) = A_2(e^{-x}\cos x - \sin x\, e^{-x})$$

Using $x'(0) = 1$,

$$1 = A_2[(1)(1) - (0)(1)]$$
$$A_2 = 1$$

The solution is

$$x(t) = \boxed{e^{-t}\sin t}$$

(b) With no damping, the differential equation would be
$$x'' + 2x = 0$$
This has a solution of $x = \sin\sqrt{2}\,t$, so

$$\omega_{\text{natural}} = \boxed{\sqrt{2}}$$

(c)
$$x(t) = e^{-t}\sin t$$
$$x'(t) = e^{-t}\cos t - \sin t\, e^{-t}$$
$$= e^{-t}(\cos t - \sin t)$$

For x to be maximum, $x'(t) = 0$. Since e^{-t} is not 0 unless t is very large, $\cos t - \sin t$ must be 0. This occurs at $t = 0.785$ radians, so

$$x(0.785) = e^{-0.785}\sin(0.785)$$
$$= \boxed{0.322}$$

(d) Use the Laplace transform method.

$$x'' + 2x' + 2x = \sin t$$
$$\mathcal{L}(x'') + 2\mathcal{L}(x') + 2\mathcal{L}(x) = \mathcal{L}(\sin t)$$
$$s^2\mathcal{L}(x) - 1 + 2s\mathcal{L}(x) + 2\mathcal{L}(x) = \frac{1}{s^2 + 1}$$
$$\mathcal{L}(x)(s^2 + 2s + 2) - 1 = \frac{1}{s^2 + 1}$$

$$\mathcal{L}(x) = \frac{1}{s^2 + 2s + 2} + \frac{1}{(s^2 + 1)(s^2 + 2s + 2)}$$

$$= \frac{1}{(s+1)^2 + 1} + \frac{1}{(s^2 + 1)(s^2 + 2s + 2)}$$

Use partial fractions to expand the second term.

$$\frac{1}{(s^2 + 1)(s^2 + 2s + 2)} = \frac{A_1 + B_1 s}{s^2 + 1} + \frac{A_2 + B_2 s}{s^2 + 2s + 2}$$

Cross multiplying,

$$= \frac{\begin{array}{c} A_1 s^2 + 2A_1 s + 2A_1 + B_1 s^3 + 2B_1 s^2 \\ + 2B_1 s + A_2 s^2 + A_2 + B_2 s^3 + B_2 s \end{array}}{(s^2 + 1)(s^2 + 2s + 2)}$$

$$= \frac{\begin{array}{c} s^3 (B_1 + B_2) + s^2 (A_1 + A_2 + 2B_1) \\ + s(2A_1 + 2B_1 + B_2) + 2A_1 + A_2 \end{array}}{(s^2 + 1)(s^2 + 2s + 2)}$$

Compare numerators to obtain the following four simultaneous equations.

$$\begin{array}{rcrcrcr} & & & & B_1 & + & B_2 & = & 0 \\ A_1 & + & A_2 & + & 2B_1 & & & = & 0 \\ 2A_1 & & & + & 2B_1 & + & B_2 & = & 0 \\ 2A_1 & + & A_2 & & & & & = & 1 \end{array}$$

Use Cramer's rule to find A_1.

$$A_1 = \frac{\begin{vmatrix} 0 & 0 & 1 & 1 \\ 0 & 1 & 2 & 0 \\ 0 & 0 & 2 & 1 \\ 1 & 1 & 0 & 0 \end{vmatrix}}{\begin{vmatrix} 0 & 0 & 1 & 1 \\ 1 & 1 & 2 & 0 \\ 2 & 0 & 2 & 1 \\ 2 & 1 & 0 & 0 \end{vmatrix}} = \frac{-1}{-5} = \frac{1}{5}$$

The rest of the coefficients are found similarly.

$$A_1 = \tfrac{1}{5}$$
$$A_2 = \tfrac{3}{5}$$
$$B_1 = -\tfrac{2}{5}$$
$$B_2 = \tfrac{2}{5}$$

Then,

$$\mathcal{L}(x) = \frac{1}{(s+1)^2 + 1} + \frac{\tfrac{1}{5}}{s^2 + 1} + \frac{-\tfrac{2}{5}s}{s^2 + 1}$$

$$+ \frac{\tfrac{3}{5}}{s^2 + 2s + 2} + \frac{\tfrac{2}{5}s}{s^2 + 2s + 2}$$

Taking the inverse transform,

$$x(t) = \mathcal{L}^{-1} \{\mathcal{L}(x)\}$$
$$= e^{-t} \sin t + \tfrac{1}{5} \sin t - \tfrac{2}{5} \cos t + \tfrac{3}{5} e^{-t} \sin t$$
$$+ \left(\tfrac{2}{5}\right) (e^{-t} \cos t - e^{-t} \sin t)$$

$$\boxed{= \tfrac{6}{5} e^{-t} \sin t + \tfrac{2}{5} e^{-t} \cos t + \tfrac{1}{5} \sin t - \tfrac{2}{5} \cos t}$$

Timed

1. This is a typical hypothesis test of two population means. The two populations are the original population the manufacturer used to determine the 1600-hr average life value and the new population the sample was taken from. The mean ($\overline{x} = 1520$ hr) of the sample and its standard deviation ($s = 120$ hr) are known, but the mean and the standard deviation of a population of average lifetimes are unknown.

(a) Assume that the average lifetime population mean and the sample mean are identical ($\overline{x} = u = 1520$ hr).

(b) The standard deviation of the average lifetime population is

$$\sigma_{\overline{x}} = \frac{s}{\sqrt{n}} = \frac{120}{\sqrt{100}} = 12$$

The manufacturer can be reasonably sure that the claim of a 1600-hr average life is justified if the average test life is near 1600 hr. "Reasonably sure" must be evaluated based on an acceptable probability of being incorrect. If the manufacturer is willing to be wrong with a 5% probability, then a 95% confidence level is required.

Since the direction of bias is known, a one-tailed test is required. To determine if the mean has shifted downward, test that 1600 hr is within the 95% limit of a distribution with a mean of 1520 hr and a standard deviation of 12 hr. From a standard normal table, 5% of a standard normal population is outside of $z = 1.645$. Therefore, the 95% confidence limit is

$$1520 + (1.645)(12) \approx 1540$$

The manufacturer can be 95% certain that the average lifetime of the bearings is less than 1600 hr (since 1600 is not between 1520 and 1540).

If the manufacturer is willing to be wrong with a probability of only 1%, then a 99% confidence limit is required. From the normal table, $z = 2.33$, and the 99% confidence limit is

$$1520 + (2.33)(12) \approx 1548$$

The manufacturer can be 99% certain that the average bearing life is less than 1600 hr.

2.

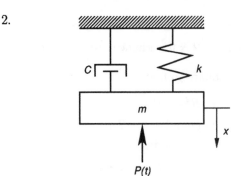

$$m = \frac{8.0\,\text{lbm}}{32.2\,\dfrac{\text{ft}}{\text{sec}^2}} = 0.25\,\text{slugs}$$

$$C = 0.50\,\text{lbf-sec/ft}$$

$$k = \frac{(8.0\,\text{lbf})\left(12\,\dfrac{\text{in}}{\text{ft}}\right)}{5.9\,\text{in}} = 16.27\,\text{lbf/ft}$$

The differential equation is

$$\left(\frac{m}{g_c}\right)\left(\frac{d^2x}{dt^2}\right) = -kx - C\left(\frac{dx}{dt}\right) + P(t)$$

$$0.25'' + 0.50x' + 16.27x = 4\cos(2t)$$

$$x'' + 2x' + 65x = 16\cos(2t)$$

Initial conditions are

$$x_0 = 0$$

$$x_0' = 0$$

Taking the Laplace transform of both sides,

$$\mathcal{L}(x'') + \mathcal{L}(2x') + \mathcal{L}(65x) = \mathcal{L}[16\cos(2t)]$$

$$s^2\mathcal{L}(x) - sx_0 - x_0' + 2s\mathcal{L}(x) - 2x_0 + 65\mathcal{L}(x) = (16)\left(\frac{s}{s^2+4}\right)$$

$$\mathcal{L}(x) = \frac{16s}{(s^2+4)(s^2+2s+65)}$$

Use partial fractions.

$$\mathcal{L}(x) = \frac{16s}{(s^2+4)(s^2+2s+65)}$$

$$= \frac{As+B}{s^2+4} + \frac{Cs+D}{s^2+2s+65}$$

$$16s = As^2 + Bs^2 + 2As^2 + 2Bs + 65As + 65B$$
$$+ Cs^3 + Ds^2 + 4Cs + 4D$$

Then,

$$A + C = 0 \qquad C = -A = -\frac{61}{8}B$$

$$B + 2A + D = 0 \quad B + 2A - \frac{65}{4}B = 0 \rightarrow A = \frac{61}{8}B$$

$$65B + 4D = 0 \qquad D = -\frac{65}{4}B$$

$$2B + 65A + 4C = 16$$

$$2B + (65)\left(\frac{61}{8}B\right) + (4)\left(-\frac{61}{8}B\right) = 16$$

$$B = 0.0342521$$
$$A = 0.2611721$$
$$C = -0.2611721$$
$$D = -0.5565962$$

Now,

$$\mathcal{L}(x) = \frac{0.2611721s + 0.0342521}{s^2+4} - \frac{0.2611721s + 0.5566}{(s+1)^2 + 8^2}$$

$$= (0.26)\left(\frac{s}{s^2+2^2}\right) + (0.017)\left(\frac{2}{s^2+2^2}\right)$$

$$- (0.26)\left(\frac{s-(-1)}{[s-(-1)]^2+8^2} + \frac{1.1311472}{[s-(-1)]^2+8^2}\right)$$

Take the inverse transform, giving

$$\boxed{\begin{aligned} x(t) &= 0.26\cos(2t) + 0.017\sin(2t) \\ &\quad - (0.26)\left[e^{-t}\cos(8t) + 0.14e^{-t}\sin(8t)\right] \end{aligned}}$$

3. $\bar{x} = \dfrac{\sum x_i}{n}$

$$= \frac{1249.529 + 1249.494 + 1249.384 + 1249.348}{4}$$

$$= 1249.4388$$

Since this is a small sample ($n < 50$), use the sample standard deviation.

$$s = \sqrt{\frac{1249.529^2 + 1249.494^2 + 1249.384^2 + 1249.348^2 - \frac{4997.755^2}{4}}{3}}$$

$$= 0.086487$$

From a standard normal table, 90% fall within $1.645s$ of $\bar{x}$, or

$$1249.4388 \pm (1.645)(0.086487) = 1249.4388 \pm 0.1423$$

$$= 1249.5811, 1249.2965$$

(a) By observation, all points fall within this range, so all are acceptable.

(b) None

(c) To be unacceptable, readings must be outside of $1.645s$ for 90% confidence.

(d) The unbiased estimate is $\bar{x} = 1249.4388$.

(e) At 90%, error $= 1.645s = 0.1423$.

(f) If the surveying crew places a marker, measures a distance x, places a second marker, then measures the same distance x back to the original marker, the ending point should coincide with the original marker. If, due to measurement errors, the ending and starting points do not coincide, the difference is the *closure error*.

In this example, the survey crew moves around the four sides of a square, so there are two measurements in the x-direction and two measurements in the y-direction. If the errors, E_1 and E_2, are known for two measurements, x_1 and x_2, the error associated with the sum or difference, $x_1 \pm x_2$, is

$$E_{x_1 \pm x_2} = \sqrt{E_1^2 + E_2^2}$$

In this case, the error in the x-direction, E_x, is

$$E_x = \sqrt{(0.1422)^2 + (0.1422)^2} = 0.2011$$

The error in the y-direction, E_y, is calculated the same way and is also 0.2011. E_x and E_y are combined by the Pythagorean theorem to yield

$$E_{\text{closure}} = \sqrt{(0.2011)^2 + (0.2011)^2}$$

$$= \boxed{0.2844}$$

(g) In surveying, error may be expressed as a fraction of one or more legs of the traverse. Assume that the total of all four legs is to be used as the basis.

$$\frac{0.2844}{(4)(1249)} = \boxed{\frac{1}{17{,}567}}$$

(h) In surveying, a second-order error is smaller than 1/10,000. 1/17,567 is smaller than a second-order error; therefore, this error is within the second order of accuracy.

(i) An experiment is said to be *accurate* if it is unaffected by experimental error. *Precision* is concerned with the repeatability of the experimental results. If an experiment is repeated with identical results, the experiment is said to be precise. However, it is possible to have a highly precise experiment with a large bias.

(j) A *systematic error* is one that is always present and is unchanged in size and direction. For example, if a steel tape is 0.02 ft short, it introduces a systematic error.

4. This is a fluid mixture problem. The water mass in the lagoon is

$$m_i = \left(\frac{\pi}{4}\right)(120\,\text{ft})^2(10\,\text{ft})\left(62.4\,\frac{\text{lbm}}{\text{ft}^3}\right) = 7.06 \times 10^6\,\text{lbm}$$

The flow rate is

$$\phi(t) = \left(30\,\frac{\text{gal}}{\text{min}}\right)\left(8.345\,\frac{\text{lbm}}{\text{gal}}\right)$$

$$= 250.4\,\text{lbm/min}$$

The mass of chemicals in the lagoon when $C = 1$ ppb is

$$m_f = (1 \times 10^{-9})(7.06 \times 10^6)$$

$$= 7.06 \times 10^{-3}\,\text{lbm}$$

$$S_{\text{in}}(t) = 0$$

$$S_{\text{out}}(t) = C(t)\phi(t)$$

$$= \frac{m(t)\left(250.4\,\dfrac{\text{lbm}}{\text{min}}\right)}{7.06 \times 10^6\,\text{lbm}}$$

$$= 3.55 \times 10^{-5}\,m(t)$$

The differential equation is

$$m'(t) = S_{\text{in}}(t) - S_{\text{out}}(t)$$

$$= -3.55 \times 10^{-5}\,m(t)$$

$$m'(t) + 3.55 \times 10^{-5}\,m(t) = 0$$

The characteristic equation is

$$R + 3.55 \times 10^{-5} = 0$$

$$R = -3.55 \times 10^{-5} = 0$$

The solution to the differential equation is

$$m(t) = Ae^{-3.55 \times 10^{-5}t}$$

Since $m = 90$ lbm when $t = 0$, $A = 90$ lbm.

$$m(t) = 90e^{-3.55 \times 10^{-5}t}$$

Solve for t when $m(t) = 7.06 \times 10^{-3}$.

$$7.06 \times 10^{-3} = 90e^{-3.55 \times 10^{-5}t}$$

$$\frac{7.06 \times 10^{-3}}{90} = e^{-3.55 \times 10^{-5}t}$$

Taking the natural log,

$$-9.45 = -3.55 \times 10^{-5}t$$

$$t = 2.66 \times 10^5\,\text{min}$$

$$= \boxed{185\,\text{days}}$$

ENGINEERING ECONOMIC ANALYSIS

Warmups

1. $F = (\$1000)(F/P, 6\%, 10)$

$= (\$1000)(1.7908) = \boxed{\$1790.80}$

2. $P = (\$2000)(P/F, 6\%, 4)$

$= (\$2000)(0.7921) = \boxed{\$1584.20}$

3. $P = (\$2000)(P/F, 6\%, 20)$

$= (\$2000)(0.3118) = \boxed{\$623.60}$

4. $\$500 = A(P/A, 6\%, 7)$

$= A(5.5824)$

$A = \dfrac{\$500}{5.5824} = \boxed{\$89.57}$

5. $F = (\$50)(F/A, 6\%, 10)$

$= (\$50)(13.1808) = \boxed{\$659.04}$

6. Each year is independent.

$\dfrac{\$200}{1.06} = \boxed{\$188.68}$

7. $\$2000 = A(F/A, 6\%, 5)$

$= A(5.6371)$

$A = \dfrac{\$2000}{5.6371} = \boxed{\$354.79}$

Alternate solution:

$\$2000 = A\big[(F/P, 6\%, 4) + (F/A, 6\%, 4)\big]$

8. $F = (\$100)[(F/P, 6\%, 10)$

$+ (F/P, 6\%, 8) + (F/P, 6\%, 6)]$

$= (\$100)(1.7908 + 1.5938 + 1.4185)$

$= \boxed{\$480.31}$

9. $r = 0.06$

$\phi = \dfrac{0.06}{12} = 0.005$

$n = (5)(12) = 60$

$F = (\$500)(1.005)^{60} = \boxed{\$674.43}$

10. $\$120 = (\$80)(F/P, i, 7)$

$(F/P, i, 7) = \dfrac{\$120}{\$80} = 1.5$

By searching the tables,

$i = \boxed{6\%}$

Concentrates

1. $\text{EUAC} = (\$17,000 + \$5000)(A/P, 6\%, 5)$

$- (\$14,000 + \$2500)(A/F, 6\%, 5) + \$200$

$= (\$22,000)(0.2374) - (\$16,500)(0.1774)$

$+ \$200$

$= \boxed{\$2495.70}$

2. Assume that the bridge will be there forever. To find the cost of keeping the old bridge, the generally accepted method is to consider the salvage value as a benefit lost (cost).

$\text{EUAC} = (\$9000 + \$13,000)(A/P, 8\%, 20)$

$- (\$10,000)(A/F, 8\%, 20) + \$500$

$= (\$22,000)(0.1019) - (\$10,000)(0.0219)$

$+ \$500$

$= \$2522.80$

Find the cost of replacing the old bridge.

$\text{EUAC} = (\$40,000)(A/P, 8\%, 25)$

$- (\$15,000)(A/F, 8\%, 25) + \$100$

$= (\$40,000)(0.0937) - (\$15,000)(0.0137) + \$100$

$= \$3642.50$

$\boxed{\text{Keep the old bridge.}}$

3. $D = \dfrac{\$150{,}000}{15} = \$10{,}000$

$$0 = \$150{,}000$$
$$+ (\$32{,}000)(1 - 0.48)(P/A, i, 15)$$
$$- (\$7530)(1 - 0.48)(P/A, i, 15)$$
$$+ (\$10{,}000)(0.48)(P/A, i, 15)$$
$$\$150{,}000 = (\$16{,}640 - \$3915.60 + \$4800)$$
$$\times (P/A, i, 15)$$
$$(P/A, i, 15) = \dfrac{\$150{,}000}{\$17{,}524.40} = 8.5595$$

By searching the tables, $i = \boxed{8\%}$

4. (a) $\dfrac{\$1{,}500{,}000 - \$300{,}000}{\$1{,}000{,}000} = \boxed{1.2}$

  (b) $\$1{,}500{,}000 - \$300{,}000 - \$1{,}000{,}000 = \boxed{\$200{,}000}$

5. The annual rent is

$$(12)(\$75) = \$900$$
$$F = (\$14{,}000 + \$1000)(F/P, 10\%, 10)$$
$$+ (\$150 + \$250 - \$900)(F/A, 10\%, 10)$$
$$= (\$15{,}000)(2.5937) - (\$500)(15.9374)$$
$$= \boxed{\$30{,}936.80}$$

6. $\$2000 = (\$89.30)(P/A, i, 30)$

$$(P/A, i, 30) = \dfrac{\$2000}{\$89.30} = 22.396$$
$$i = 2\% \text{ per month}$$
$$= (1.02)^{12} - 1 = \boxed{0.2682 \text{ or } 26.82\%}$$

7. (a) SL:

$$D = \dfrac{\$500{,}000 - \$100{,}000}{25} = \boxed{\$16{,}000}$$

  (b) SOYD:

$$T = \left(\tfrac{1}{2}\right)(25)(26) = 325$$
$$D_1 = \left(\dfrac{25}{325}\right)(\$500{,}000 - \$100{,}000) = \boxed{\$30{,}769}$$
$$D_2 = \left(\dfrac{24}{325}\right)(\$400{,}000) = \boxed{\$29{,}538}$$
$$D_3 = \left(\dfrac{23}{325}\right)(\$400{,}000) = \boxed{\$28{,}308}$$

(c) DDB:

$$D_1 = \left(\dfrac{2}{25}\right)(\$500{,}000) = \boxed{\$40{,}000}$$
$$D_2 = \left(\dfrac{2}{25}\right)(\$500{,}000 - \$40{,}000) = \boxed{\$36{,}800}$$
$$D_3 = \left(\dfrac{2}{25}\right)(\$500{,}000 - \$40{,}000 - \$36{,}800)$$
$$= \boxed{\$33{,}856}$$

8. $P = -\$12{,}000 + (\$2000)(P/F, 10\%, 10)$

$$- (\$1000)(P/A, 10\%, 10)$$
$$- (\$200)(P/G, 10\%, 10)$$
$$= -\$12{,}000 + (\$2000)(0.3855)$$
$$- (\$1000)(6.1446) - (\$200)(22.8913)$$
$$= \boxed{-\$21{,}951.86}$$

$$\text{EUAC} = (\$21{,}951.86)(A/P, 10\%, 10)$$
$$= (\$21{,}951.86)(0.1627)$$
$$= \boxed{\$3571.57}$$

9. Assume that the probability of failure in any of the n years is $1/n$.

$$\text{EUAC}(9) = (\$1500)(A/P, 6\%, 20)$$
$$+ \left(\dfrac{1}{9}\right)(0.35)(\$1500)$$
$$+ (0.04)(\$1500)$$
$$= (\$1500)\left[0.0872 + (0.35)\left(\dfrac{1}{9}\right) + 0.04\right]$$
$$= \$249.13$$
$$\text{EUAC}(14) = (\$1600)\left[0.1272 + (0.35)\left(\dfrac{1}{14}\right)\right]$$
$$= \$243.52$$
$$\text{EUAC}(30) = (\$1750)\left[0.1272 + (0.35)\left(\dfrac{1}{30}\right)\right]$$
$$= \$243.02$$
$$\text{EUAC}(52) = (\$1900)\left[0.1272 + (0.35)\left(\dfrac{1}{52}\right)\right]$$
$$= \$254.47$$
$$\text{EUAC}(86) = (\$2100)\left[0.1272 + (0.35)\left(\dfrac{1}{86}\right)\right]$$
$$= \$275.67$$

$\boxed{\text{Choose the 30-year pipe.}}$

10. $\text{EUAC}(7) = (0.15)(\$25{,}000) = \$3750$

$\text{EUAC}(8) = (\$15{,}000)(A/P, 10\%, 20)$
$+ (0.10)(\$25{,}000)$
$= (\$15{,}000)(0.1175) + (0.10)(\$25{,}000)$
$= \$4262.50$

$\text{EUAC}(9) = (\$20{,}000)(0.1175) + (0.07)(\$25{,}000)$
$= \$4100$

$\text{EUAC}(10) = (\$30{,}000)(0.1175) + (0.03)(\$25{,}000)$
$= \$4275$

$\boxed{\text{It is cheapest to do nothing.}}$

Timed

1. $\text{EUAC}(1) = (\$10{,}000)(A/P, 20\%, 1) + (\$2000)$
$- (\$8000)(A/F, 20\%, 1)$
$= (\$10{,}000)(1.2000) + \$2000$
$- (\$8000)(1.0000)$
$= \$6000$

$\text{EUAC}(2) = (\$10{,}000)(A/P, 20\%, 2) + \$2000$
$+ (\$1000)(A/G, 20\%, 2)$
$- (\$7000)(A/F, 20\%, 2)$
$= (\$10{,}000)(0.6545) + \$2000$
$+ (\$1000)(0.4545)$
$- (\$7000)(0.4545)$
$= \$5818$

$\text{EUAC}(3) = (\$10{,}000)(A/P, 20\%, 3)$
$+ \$2000 + (\$1000)(A/G, 20\%, 3)$
$- (\$6000)(A/F, 20\%, 3)$
$= (\$10{,}000)(0.4747) + \$2000$
$+ (\$1000)(0.8791)$
$- (\$6000)(0.2747)$
$= \$5977.90$

$\text{EUAC}(4) = (\$10{,}000)(A/P, 20\%, 4)$
$+ \$2000 + (\$1000)(A/G, 20\%, 4)$
$- (\$5000)(A/F, 20\%, 4)$
$= (\$10{,}000)(0.3863) + \$2000$
$+ (\$1000)(1.2742)$
$- (\$5000)(0.1863)$
$= \$6205.70$

$\text{EUAC}(5) = (\$10{,}000)(A/P, 20\%, 5)$
$+ \$2000 + (\$1000)(A/G, 20\%, 5)$
$- (\$4000)(A/F, 20\%, 5)$
$= (\$10{,}000)(0.3344) + \$2000$
$+ (\$1000)(1.6405) - (\$4000)(0.1344)$
$= \$6446.90$

(a) $\boxed{\text{Sell at the end of the second year.}}$

(b) From Eq. 2.8, the cost of owning the structure one more year is

$\$6000 + (0.20)(\$5000) + (\$5000 - \$4000) = \boxed{\$8000}$

2. The man should charge his company for only the costs due to the business travel.

insurance: $\$300 - \$200 = \$100$

maintenance: $\$200 - \$150 = \$50$

salvage value
reduction: $\$1000 - \$500 = \$500$
$(\$500)(A/F, 10\%, 5) = (\$500)(0.1638)$
$= \$81.90$

gasoline: $\dfrac{(\$5000)(0.60)}{15} = \$200$

$$\text{EUAC per mile} = \frac{\$100 + \$50 + \$81.9 + \$200}{5000}$$
$$= \$0.0864 \text{ per mile}$$

(a) $\boxed{\$0.10 > \$0.0864, \text{ so the reimbursement is adequate.}}$

(b) $0.10x = (\$5000)(A/P, 10\%, 5) + \$250 + \$200$
$- (\$800)(A/F, 10\%, 5) + \left(\dfrac{x}{15}\right)(0.60)$
$= (\$5000)(0.2638) + \$250 + \$200$
$- (\$800)(0.1638) + 0.04x$

$0.10x = 1637.96 + 0.04x$

$0.06x = 1637.96$

$x = \boxed{27{,}299 \text{ mi}}$

PROFESSIONAL PUBLICATIONS, INC. ● Belmont, CA

3. Use EUAC since the lives are different.

$$P\{A\} = -\$80,000 - (\$5000)(P/F, 10\%, 10)$$
$$+ (\$7000)(P/F, 10\%, 20)$$
$$- (\$2000)(P/A, 10\%, 20)$$
$$+ (\$500)(P/A, 10\%, 10)$$
$$+ (\$500)(P/A, 10\%, 5)$$
$$= -\$80,000 - (\$5000)(0.3855)$$
$$+ (\$7000)(0.1486)$$
$$- (\$2000)(8.5136)$$
$$+ (\$500)(6.1446 + 3.7908)$$
$$= -\$92,947$$

$$EUAC\{A\} = (\$92,947)(A/P, 10\%, 20)$$
$$= (\$92,947)(0.1175)$$
$$= \$10,921$$

$$P\{B\} = -\$35,000 - (\$4000)(P/A, 10\%, 10)$$
$$+ (\$1000)(P/A, 10\%, 5)$$
$$= -\$35,000 - (\$4000)(6.1446)$$
$$+ (\$1000)(3.7908)$$
$$= -\$55,788$$

$$EUAC\{B\} = (\$55,788)(A/P, 10\%, 10)$$
$$= (\$55,788)(0.1627)$$
$$= \$9077$$

$$\boxed{B \text{ has the lowest cost.}}$$

4. Let x be the number of miles driven per year. The EUAC for both alternatives is

$$EUAC(A) = 0.15x$$
$$EUAC(B) = 0.04x + \$500 + (\$5000)(A/P, 10\%, 3)$$
$$- (\$1200)(A/F, 10\%, 3)$$
$$= 0.04x + \$500 + (\$5000)(0.4021)$$
$$- (\$1200)(0.3021)$$
$$= 0.04x + 2148$$

Setting these equal,

$$0.15x = 0.04x + 2148$$

$$x = \boxed{19,527 \text{ mi}}$$

5. (a) The operating costs per year are

$$A: (\$10.50)(365)(24) = \$91,980$$
$$B: (\$8)(365)(24) = \$70,080$$

$$EUAC\{A\} = (\$13,000 + \$10,000)(A/P, 7\%, 10)$$
$$+ \$91,980 - (\$5000)(A/F, 7\%, 10)$$
$$= (\$23,000)(0.1424) + \$91,980$$
$$- (\$5000)(0.0724)$$
$$= \$94,893$$

The cost per ton-year is

$$cost(A) = \frac{\$94,893}{50} = \boxed{\$1898}$$

The present worth of B, exclusive of operating costs, is

$$P\{B\} = -\$8000$$
$$- (\$7000 + \$2200 - \$2000)(P/F, 7\%, 5)$$
$$+ (\$2000)(P/F, 7\%, 10)$$
$$= -(\$8000) - (\$7200)(0.7130)$$
$$+ (\$2000)(0.5083)$$
$$= -\$12,117$$

$$EUAC\{B\} = \$70,080 + (\$12,117)(A/P, 7\%, 10)$$
$$= \$70,080 + (\$12,117)(0.1424)$$
$$= \$71,805$$

The cost per ton-year for B is

$$cost(B) = \frac{\$71,805}{20} = \boxed{\$3590}$$

(b)

tons/yr	cost of using A	cost of using B	cheapest
0–20	$94,893 (1x)	$71,805 (1x)	B
20–40	$94,893 (1x)	$143,610 (2x)	A
40–50	$94,893 (1x)	$215,415 (3x)	A
50–60	$189,786 (2x)	$215,415 (3x)	A
60–80	$189,786 (2x)	$287,220 (4x)	A

6. first-year cost $= (1)(\$37,440) = \$37,440$

gradient $= (0.1)(\$37,440) = \$3744$

(a) $EUAC = (\$60,000)(A/P, 7\%, 20) + \$37,440$
$$+ (\$3744)(A/G, 7\%, 20)$$
$$- (\$10,000)(A/F, 7\%, 20)$$
$$= (\$60,000)(0.0944) + \$37,440$$
$$+ (\$3744)(7.3163) - (\$10,000)(0.0244)$$
$$= \$70,252.23$$

For 80,000 passengers, the required fare is

$$\frac{\$70,252.23}{80,000} = \boxed{\$0.878 \text{ per passenger trip}}$$

(b) $\$0.878 = \$0.35 + (\text{gradient})(A/G, 7\%, 20)$

$$\text{gradient} = \frac{\$0.878 - \$0.35}{7.3163}$$

$$= \boxed{\$0.072 \ (7.2 \ \text{¢}) \ \text{increase per year}}$$

(c) As in part (b), the subsidy should be

$$\text{subsidy} = \text{cost} - \text{revenue}$$

$$P = \$0.878 - [\$0.35 + (0.05)(A/G, 7\%, 20)]$$

$$= \$0.878 - \$0.35 - (0.05)(7.3163)$$

$$= \boxed{\$0.162}$$

FLUID STATICS AND DYNAMICS

Warmups

1. For air at 14.7 psia and 80°F,

$$\mu = 3.85 \times 10^{-7} \text{ lbf-sec/ft}^2$$

[independent of pressure]

$$\rho = \frac{p}{RT} = \frac{(70)(144)}{(53.3)(80+460)}$$

$$= 0.350 \text{ lbm/ft}^3$$

$$\nu = \frac{\mu g_c}{\rho}$$

$$= \frac{\left(3.85 \times 10^{-7} \dfrac{\text{lbf-sec}}{\text{ft}^2}\right)\left(32.2 \dfrac{\text{lbm-ft}}{\text{lbf-sec}^2}\right)}{0.350 \dfrac{\text{lbm}}{\text{ft}^3}}$$

$$= \boxed{3.54 \times 10^{-5} \text{ ft}^2/\text{sec}}$$

2. $$p = 14.7 - 8.7 = \boxed{6.0 \text{ psia}}$$

3. $$p_1 - p_2 = (7)(0.491) - (7)(0.0361)$$

$$= \boxed{3.184 \text{ psi}}$$

4. The perimeter is

$$p = \pi d = \pi(10)$$

$$= 31.42 \text{ in}$$

Assume an elliptical cross section.

$$b = \left(\tfrac{1}{2}\right)(7.2) = 3.6$$

$$2\pi\sqrt{\left(\tfrac{1}{2}\right)[a^2 + (3.6)^2]} = 31.42$$

$$a = 6.09 \text{ in}$$

The area in flow is

$$A = \tfrac{1}{2}\pi ab = \left(\tfrac{1}{2}\right)\pi(6.09)(3.6)$$

$$= 34.44 \text{ in}^2$$

$$r_h = \frac{34.44}{\left(\tfrac{1}{2}\right)(31.42)}$$

$$= \boxed{2.19 \text{ in}}$$

5. Assume schedule-40 pipe and 70°F water.

$$D_i = 0.5054 \text{ ft}$$

$$A_i = 0.2006 \text{ ft}^2$$

$$v = \frac{1.5 \dfrac{\text{ft}^3}{\text{sec}}}{0.2006 \text{ ft}^2}$$

$$= 7.478 \text{ ft/sec}$$

At 70°F,

$$\nu = 1.059 \times 10^{-5} \text{ ft}^2/\text{sec}$$

$$N_{\text{Re}} = \frac{Dv}{\nu} = \frac{(0.5054 \text{ ft})\left(7.478 \dfrac{\text{ft}}{\text{sec}}\right)}{1.059 \times 10^{-5} \dfrac{\text{ft}^2}{\text{sec}}}$$

$$= \boxed{3.57 \times 10^5}$$

6. $$\epsilon = 0.0002 \text{ ft}$$

$$\frac{\epsilon}{D} = \frac{0.0002}{0.5054} = 0.0004$$

From the Moody friction factor chart,

$$f \approx 0.0175$$

$$h_f = \frac{(0.0175)(1200 \text{ ft})\left(7.478 \dfrac{\text{ft}}{\text{sec}}\right)^2}{(0.5054 \text{ ft})(2)\left(32.2 \dfrac{\text{ft}}{\text{sec}^2}\right)}$$

$$= \boxed{36.1 \text{ ft}}$$

7. Assume schedule-40 steel pipe.

At A,

$$D_{\text{A}} = 0.5054 \text{ ft}$$

$$A_{\text{A}} = 0.2006 \text{ ft}^2$$

$$v_{\text{A}} = \frac{5 \dfrac{\text{ft}^3}{\text{sec}}}{0.2006 \text{ ft}^2}$$

$$= 24.925 \text{ ft/sec}$$

$$p_{\text{A}} = 10 \text{ psia}$$

$$= 1440 \text{ psf}$$

$$z_{\text{A}} = 0$$

At B,

$$D_B = 1.4063 \text{ ft}$$
$$A_B = 1.5533 \text{ ft}^2$$
$$v_B = \frac{5}{1.5533}$$
$$= 3.219 \text{ ft/sec}$$
$$p_B = 7 \text{ psia}$$
$$= 1008 \text{ psf}$$
$$z_B = 15$$

Disregard the minor losses and assume 70°F water, so $\rho = 62.3 \text{ lbm/ft}^3$. The total heads at A and B are

$$(TH)_A = \frac{1440}{62.3} + \frac{(24.925)^2}{(2)(32.2)} + 0$$
$$= 32.76 \text{ ft}$$
$$(TH)_B = \frac{1008}{62.3} + \frac{(3.219)^2}{(2)(32.2)} + 15$$
$$= 31.3 \text{ ft}$$

$$\boxed{\text{The flow is from A to B.}}$$

8.
$$r = \frac{v_0^2 \sin 2\phi}{g}$$
$$= \frac{(50)^2 \sin [(2)(45)]}{32.2}$$
$$= \boxed{77.64 \text{ ft}}$$

9. The mass of water at 70°F is

$$\left(100 \frac{\text{ft}^3}{\text{sec}}\right)\left(62.4 \frac{\text{lbm}}{\text{ft}^3}\right) = 6240 \text{ lbm/sec}$$

The energy loss in feet of water is

$$\left[\frac{\left(30 \frac{\text{lbf}}{\text{in}^2} + 5 \frac{\text{lbf}}{\text{in}^2}\right)\left(144 \frac{\text{in}^2}{\text{ft}^2}\right)}{62.4 \frac{\text{lbm}}{\text{ft}^3}}\right]\left(\frac{g_c}{g}\right) = 80.77 \text{ ft}$$

The horsepower is

$$\left[\frac{\left(6240 \frac{\text{lbm}}{\text{sec}}\right)(80.77 \text{ ft})}{550 \frac{\text{ft-lbf}}{\text{hp-sec}}}\right]\left(\frac{g}{g_c}\right) = \boxed{916.4 \text{ hp}}$$

10. Assume schedule-40 pipe and 70°F water.

$$D_1 = 7.981 \text{ in}$$
$$D_2 = 6.065 \text{ in}$$

The mass flow is

$$A_2 v_2 \rho = \left(\frac{\pi}{4}\right)\left(\frac{6.065}{12}\right)^2 (12)(62.4)$$
$$= 150.2 \text{ lbm/sec}$$

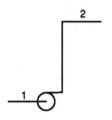

The inlet pressure is

$$(14.7 - 5)(144) = 1396.8 \text{ lbf/ft}^2$$

The head added by the pump is

$$h_A = \frac{(0.70)(20)(550)}{150.2}$$
$$= 51.26 \text{ ft}$$

At 1,

$$p_1 = 1396.8 \text{ lbf/ft}^2$$
$$v_1 = (12)\left(\frac{6.065}{7.981}\right)^2$$
$$= 6.93 \text{ ft/sec}$$
$$z_1 = 0$$

At 2,

$$p_2 = (14.7)(144)$$
$$= 2116.8 \text{ lbf/ft}^2$$
$$v_2 = 12 \text{ ft/sec}$$

Using the Bernoulli equation,

$$\frac{1396.8}{62.4} + \frac{(6.93)^2}{(2)(32.2)} + 51.26 = \frac{2116.8}{62.4} + \frac{(12)^2}{(2)(32.2)}$$
$$+ z_2 + 10$$
$$z_2 = \boxed{28.2 \text{ ft}}$$

Concentrates

1. The buoyant force is equal to the weight of the displaced air, which has the same volume as the hydrogen.

$$V = \frac{mRT}{p}$$

$$= \frac{(10,000)(766.8)(460+56)}{(30.2)(0.491)(144)}$$

$$= 1.853 \times 10^6 \text{ ft}^3$$

The displaced air has a mass of

$$m = \frac{pV}{RT}$$

$$= \frac{(30.2)(0.491)(144)(1.85 \times 10^6)}{(53.3)(460+56)}$$

$$= 1.436 \times 10^5 \text{ lbm} \quad \begin{bmatrix} \text{equal numerically} \\ \text{to the weight} \end{bmatrix}$$

Neglecting the structural weight, the lift is

$$F_{\text{buoyant}} - w_{\text{He}} = 1.436 \times 10^5 - 10,000$$

$$= \boxed{1.336 \times 10^5 \text{ lbf}}$$

2.
$$D_i = 0.5054 \text{ ft}$$

$$A_i = 0.2006 \text{ ft}^2$$

The volume flow is

$$(500 \text{ gpm}) \left(0.00223 \, \frac{\text{ft}^3}{\text{sec-gpm}} \right) = 1.115 \text{ ft}^3/\text{sec}$$

The fluid velocity is

$$\text{v} = \frac{Q}{A} = \frac{1.115}{0.2006}$$

$$= 5.558 \text{ ft/sec}$$

At 100°F,

$$\nu = 0.739 \times 10^{-5} \text{ ft}^2/\text{sec}$$

$$N_{\text{Re}} = \frac{\text{v}D}{\nu} = \frac{(5.558)(0.5054)}{0.739 \times 10^{-5}}$$

$$= 3.8 \times 10^5$$

$$\epsilon = 0.0002 \text{ ft}$$

$$\frac{\epsilon}{D} = \frac{0.0002}{0.5054}$$

$$= 0.0004$$

From the Moody chart, $f = 0.0175$. The approximate equivalent lengths of the fittings are

$$
\begin{array}{lrl}
\text{elbows:} & 2 \times 8.9 = & 17.8 \\
\text{gate valve:} & 2 \times 3.2 = & 6.4 \\
90° \text{ valve:} & 1 \times 63.0 = & 63.0 \\
\text{backflow limiter:} & 1 \times 63.0 = & \underline{63.0} \\
& & \overline{150.2 \text{ ft}}
\end{array}
$$

The friction loss across the 300 ft is

$$L_e = 300 + 150.2$$

$$= 450.2 \text{ ft}$$

$$h_f = \frac{(0.0175)(450.2)(5.558)^2}{(2)(0.5054)(32.2)}$$

$$= 7.48 \text{ ft}$$

The pressure difference for 100°F water is

$$p = \gamma h = \rho h \left(\frac{g}{g_c} \right)$$

$$= (62.0)(7.48 + 20)$$

$$= \boxed{1703.8 \text{ psf}}$$

3. For 70°F air (assumed incompressible),

$$\nu \approx 16.15 \times 10^{-5} \text{ ft}^2/\text{sec}$$

$$N_{\text{Re}} = \frac{(60)(0.5054)}{16.15 \times 10^{-5}}$$

$$= 1.88 \times 10^5$$

Using $\epsilon/D = 0.004$,

$$f = 0.0185$$

The friction head loss is

$$h_f = \frac{(0.0185)(450.2)(60)^2}{(2)(0.5054)(32.2)}$$

$$= 921.2 \text{ ft}$$

Assuming $\rho_{\text{air}} \approx 0.075 \text{ lbm/ft}^3$, the pressure difference is

$$\Delta p = \gamma h = \rho h \left(\frac{g}{g_c} \right)$$

$$= (0.075)(921.20 + 20)$$

$$= \boxed{70.59 \text{ psf}}$$

4. $(2000 \text{ gpm}) \left(0.00223 \dfrac{\text{ft}^3}{\text{sec-gpm}} \right) = 4.46 \text{ ft}^3/\text{sec}$

Assume schedule-40 pipe.

$$D_1 = 0.9948 \text{ ft}$$
$$A_1 = 0.7773 \text{ ft}^2$$
$$D_2 = 0.6651$$
$$A_2 = 0.3474$$
$$p_1 = [14.7 - (6)(0.491)](144)$$
$$= 1692.6 \text{ psf}$$
$$p_2 = [14.7 + 20](144) + (4)(1.2)(62.4)$$
$$= 5296.3 \text{ psf}$$
$$v_1 = \frac{4.46}{0.7773}$$
$$= 5.74 \text{ ft/sec}$$
$$v_2 = \frac{4.46}{0.3474}$$
$$= 12.84 \text{ ft/sec}$$

The total heads at 1 and 2 are

$$(\text{TH})_1 = \frac{1692.6}{(62.4)(1.2)} + \frac{(5.7)^2}{(2)(32.2)}$$
$$= 23.11 \text{ ft}$$
$$(\text{TH})_2 = \frac{5296.3}{(62.4)(1.2)} + \frac{(12.84)^2}{(2)(32.2)}$$
$$= 73.29 \text{ ft}$$

The pump must add $73.29 - 23.11 = 50.18$ ft of head. The power required is

$$p = \Delta h \dot{m} \left(\frac{g}{g_c} \right)$$

$$= \left[\frac{(50.18 \text{ ft})(1.2) \left(62.4 \dfrac{\text{lbm}}{\text{ft}^3} \right) \left(4.46 \dfrac{\text{ft}^3}{\text{sec}} \right)}{550 \dfrac{\text{ft-lbf}}{\text{hp-sec}}} \right] \left(\frac{g}{g_c} \right)$$

$$= 30.47 \text{ hp}$$

The input horsepower is

$$\frac{30.47 \text{ hp}}{0.85} = \boxed{35.85 \text{ hp}}$$

5. This cannot be solved with the *Mechanical Engineering Reference Manual*. Refer to p. 2-13 of Crane's Technical Paper 410.

$$K_B = K_1 + (n - 1) \left[0.25 f_T \pi \left(\frac{r}{D} \right) + 0.5 K_1 \right]$$

Assume type-K copper tubing.

$$D_i = 2.125 - (2)(0.083) = 1.959 \text{ in}$$
$$= 0.1633 \text{ ft}$$
$$\frac{r}{D} = \frac{1}{0.1633}$$
$$= 6.124$$

From p. A-29 of Crane's for a 90° bend,

$$K_1 \approx 17.4 f$$
$$n = \text{no. 90° bends} = (4)(5)$$
$$= 20$$

Assume smooth copper pipe.

$$\epsilon = 0.000005$$
$$\frac{\epsilon}{D} = \frac{0.000005}{0.1633}$$
$$= 0.000031$$

The fully turbulent friction factor is

$$f_t = 0.0095$$
$$K_1 = (17.4)(0.0095)$$
$$= 0.165$$
$$K_B = 0.165 + (20 - 1)[(0.25)(0.0095)\pi(6.124)$$
$$+ (0.5)(0.165)]$$
$$= 2.6$$
$$h_f = K_B \left(\frac{v^2}{2g} \right) = (2.6) \left[\frac{(10)^2}{(2)(32.2)} \right]$$
$$= \boxed{4.04 \text{ ft}}$$

6. $\dot{Q} = \left(\dfrac{\pi}{4} \right) \left(\dfrac{2 \text{ in}}{12 \dfrac{\text{in}}{\text{ft}}} \right)^2 \left(40 \dfrac{\text{ft}}{\text{sec}} \right)$

$$= 0.8727 \text{ ft}^3/\text{sec}$$
$$\dot{Q}' = \left(\frac{40 - 15}{40} \right) (0.8727)$$
$$= 0.5454 \text{ ft}^3/\text{sec}$$
$$F_x = \left[\frac{-(0.5454)(62.4)}{32.2} \right] (40 - 15)(1 - \cos 60°)$$
$$= -13.2 \text{ lbf} \quad \text{[force of fluid is to the left]}$$
$$F_y = \left[\frac{-(0.5454)(62.4)}{32.2} \right] (40 - 15)(\sin 60°)$$
$$= 22.9 \text{ lbf} \quad \text{[force on fluid is upwards]}$$
$$F = \sqrt{(13.2)^2 + (22.9)^2}$$
$$= \boxed{26.4 \text{ lbf}}$$

7. For dynamic similitude,

$$N_{\text{Re,model}} = N_{\text{Re,true}}$$

$$\left(\frac{vL}{\nu}\right)_m = \left(\frac{vL}{\nu}\right)_t$$

$$\left(\frac{vL\rho}{\mu g}\right)_m = \left(\frac{vL\rho}{\mu g}\right)_t$$

$\rho = p/RT$ for ideal gases.

$$\left(\frac{vL\rho}{\mu gRT}\right)_m = \left(\frac{vL\rho}{\mu gRT}\right)_t$$

$$v_m = v_t$$

$$L_m = \frac{L_t}{20}$$

$$T_m = T_t$$

$$g = g$$

$$R = R$$

$$\mu_m = \mu_t \quad \text{[independent of pressure]}$$

$$p_m = \boxed{20p_t} \quad \text{[units depend on units of } p_t\text{]}$$

8.
$$(N_{\text{Re}})_m = (N_{\text{Re}})_t$$

$$\left(\frac{vD}{\nu}\right)_m = \left(\frac{vD}{\nu}\right)_t$$

D is the impeller (not pipe) diameter, and v is the tangential velocity.

$$v \propto (\text{rpm})(\text{diameter})$$

$$v_m \propto 2\,\text{rpm}_m$$

$$v_t \propto (1)(1000)$$

$$D_m = 2D_t$$

$$\nu_{\text{air},68°} = 16.0 \times 10^{-5} \text{ ft}^2/\text{sec}$$

$$\nu_{\text{castor oil}} \approx 1110 \times 10^{-5} \text{ ft}^2/\text{sec}$$

$$\frac{(2)(\text{rpm})_m(2)(D_t)}{16 \times 10^{-5}} = \frac{1000D_t}{1110 \times 10^{-5}}$$

$$(\text{rpm})_m = \boxed{3.6}$$

9. The specific gravity of benzene at 60°F is ≈ 0.885. The density is

$$(0.885)(62.4) = 55.2 \text{ lbm/ft}^3$$

$$= 0.0320 \text{ lbm/in}^3$$

The pressure difference is

$$\Delta p = (4)(0.491 - 0.0320) = 1.836 \text{ psi}$$

$$= 264.4 \text{ psf}$$

$$\left(\frac{A_2}{A_1}\right)^2 = \left(\frac{D_2}{D_1}\right)^4$$

$$F_{\text{va}} = \frac{1}{\sqrt{1 - \left(\frac{3.5}{8}\right)^4}}$$

$$= 1.019$$

$$A_2 = \left(\frac{\pi}{4}\right)\left(\frac{3.5}{12}\right)^2$$

$$= 0.0668 \text{ ft}^4$$

$$\dot{V} = (1.019)(0.99)(0.0668)\sqrt{\frac{(2)(32.2)(264.4)}{55.2}}$$

$$= \boxed{1.184 \text{ ft}^3/\text{sec}}$$

10. Assume schedule-40 pipe.

$$D_i = 0.9948 \text{ ft}$$

$$A_i = 0.7773 \text{ ft}^2$$

$$v_1 = \frac{Q}{A} = \frac{10}{0.7773}$$

$$= 12.87 \text{ ft/sec}$$

For 70°F water,

$$\nu = 1.059 \times 10^{-5} \text{ ft}^2/\text{sec}$$

$$N_{\text{Re}} = \frac{vD}{\nu} = \frac{(12.87)(0.9948)}{1.059 \times 10^{-5}}$$

$$= \boxed{1.21 \times 10^6}$$

$$\dot{V} = C_f A_o \sqrt{\frac{2g(p_1 - p_2)}{\rho}}$$

$$C_f A_o \sqrt{(2)(32.2)(25)} = 10$$

$$C_f A_o = 0.249$$

Both C_f and A_o depend on D_o.

Assume $D_o = 6$ in.

$$A_o = \left(\frac{\pi}{4}\right)\left(\frac{6}{12}\right)^2$$

$$= 0.1963 \text{ ft}^2$$

$$\frac{A_o}{A_1} = \frac{0.1963}{0.7773}$$

$$= 0.25$$

$$N_{Re} = 1.21 \times 10^6$$

$$C_f = 0.63$$

$$C_f A_o = (0.63)(0.1963)$$

$$= 0.124 \quad \text{[too small]}$$

Assume $D_o = 9$ in.

$$A_o = \left(\frac{\pi}{4}\right)\left(\frac{9}{12}\right)^2 = 0.442$$

$$\frac{A_o}{A_1} = \frac{0.442}{0.7773} = 0.57$$

$$C_f \approx 0.73$$

$$C_f A_o = (0.73)(0.442)$$

$$= 0.323 \quad \text{[too high]}$$

Interpolating,

$$D_o = 6 \text{ in} + (9-6)\left(\frac{0.249 - 0.124}{0.323 - 0.124}\right)$$

$$= 7.9$$

Further iterations yield

$$D_o \approx 8.1 \text{ in}$$

$$C_f A_o \approx \boxed{0.243}$$

Timed

1.

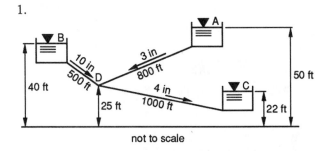

not to scale

Assume that flows from reservoirs A and B are toward D and then toward C.

$$Q_{A-D} + Q_{B-D} - Q_{D-C} = 0$$

$$A_A v_{A-D} + A_B v_{B-D} - A_C v_{D-C} = 0$$

Assume a schedule-40 pipe so that

$$A_A = 0.05134 \text{ ft}^2 \qquad D_A = 0.2557 \text{ ft}$$
$$A_B = 0.5476 \text{ ft}^2 \qquad D_B = 0.8350 \text{ ft}$$
$$A_C = 0.08841 \text{ ft}^2 \qquad D_C = 0.3355 \text{ ft}$$

$$(0.05134)v_{A-D} + (0.5476)v_{B-D} - (0.08841)v_{D-C} = 0$$
$$\text{[Eq. 1]}$$

Disregarding the velocity heads, the conservation of energy between A and D is

$$z_A = \frac{p_D}{\rho} + z_D + h_{f,A--D}$$

$$50 = \frac{p_D}{62.4} + 25 + \frac{(0.02)(800)(v_{A-D})^2}{(2)(0.2557)(32.2)}$$

or

$$v_{A-D} = \sqrt{25.73 - 0.0165 p_D} \qquad \text{[Eq. 2]}$$

Similarly, for B–D,

$$40 = \frac{p_D}{62.4} + 25 + \frac{(0.02)(500)(v_{B-D})^2}{(2)(0.8350)(32.2)}$$

or

$$v_{B-D} = \sqrt{80.66 - 0.0862 p_D} \qquad \text{[Eq. 3]}$$

For D–C,

$$22 = \frac{p_D}{62.4} + 25 - \frac{(0.02)(1000)(v_{D-C})^2}{(2)(0.3355)(32.2)}$$

or

$$v_{D-C} = \sqrt{3.24 + 0.0173 p_D} \qquad \text{[Eq. 4]}$$

Assume $p_D = 935$ psf, which is the largest it can be to keep Eqs. 2 and 3 real. Then,

$$v_{A-D} = 3.21 \text{ ft/sec}$$
$$v_{B-D} = 0.25 \text{ ft/sec}$$
$$v_{D-C} = 4.41 \text{ ft/sec}$$

From Eq. 1,

$$(0.05134)(3.21) + (0.5476)(0.25)$$
$$- (0.08841)(4.41) = -0.088$$

PROFESSIONAL PUBLICATIONS, INC. ● Belmont, CA

This is repeated using interpolation.

trial	p_D	right-hand side of Eq. 1	How was p_D chosen?
1	935	−0.088	to keep 2 and 3 real
2	900	0.745	arbitrarily
3	931.3	0.116	interpolation with trials 1 and 2
4	933.4	0.02	interpolation, 1 and 3
5	933.7	0.005	interpolation, 1 and 4

If $p_D = 933.7$ psf,

$$v_{A-D} = 3.213 \text{ ft/sec}$$

$$v_{B-D} = 0.4184 \text{ ft/sec}$$

$$v_{D-C} = 4.404 \text{ ft/sec}$$

$$\boxed{\text{Flow is out of B.}}$$

(If it had been assumed that reservoir A fed both B and C, it would have been impossible to satisfy the continuity equation.)

$$A_A v_{AD} + A_B v_{BD} - A_C v_{BC} = 0$$

2. Assume 70°F water.

$$E = 320 \times 10^3 \text{ psi}$$

$$\rho = 62.3 \text{ lbf/ft}^3$$

Although E varies considerably with the class of the cast iron, assume $E = 20 \times 10^6$ psi. The composite modulus of elasticity is

$$E = \frac{E_{\text{water}} t_{\text{pipe}} E_{\text{pipe}}}{t_{\text{pipe}} E_{\text{pipe}} + d_{\text{pipe}} E_{\text{water}}}$$

$$= \frac{(320 \times 10^3 \text{ psi})(0.75 \text{ in})(20 \times 10^6 \text{ psi})}{(0.75 \text{ in})(20 \times 10^6 \text{ psi}) + (24 \text{ in})(320 \times 10^3 \text{ psi})}$$

$$= 2.12 \times 10^5 \text{ psi}$$

$$c = \sqrt{\frac{E g_c}{\rho}}$$

$$= \sqrt{\frac{(2.12 \times 10^5)(32.2)(144)}{62.3}}$$

$$= 3972 \text{ ft/sec}$$

(The answer is 4880 ft/sec if pipe elasticity is disregarded.)

(a) The pressure increase is

$$\Delta p = \frac{\rho c \Delta v}{g_c}$$

$$= \frac{\left(62.3 \dfrac{\text{lbm}}{\text{ft}^3}\right)\left(3972 \dfrac{\text{ft}}{\text{sec}}\right)\left(6 \dfrac{\text{ft}}{\text{sec}}\right)}{32.2 \dfrac{\text{ft-lbm}}{\text{lbf-sec}^2}}$$

$$= \boxed{46{,}110 \text{ psf}}$$

(b) The maximum closure time without changing Δp is

$$t = \frac{2L}{c} = \frac{(2)(500 \text{ ft})}{3972 \dfrac{\text{ft}}{\text{sec}}}$$

$$= \boxed{0.252 \text{ sec}}$$

3. (a) $$D = \left(\tfrac{1}{2}\right)\left(\frac{C_D A \rho v^2}{g_c}\right)$$

For air at 68°F,

$$\rho = 0.0752 \text{ lbm/ft}^3$$

$$v = \frac{(55)(5280)}{3600}$$

$$= 80.67 \text{ ft/sec}$$

$$D = \frac{\left(\tfrac{1}{2}\right)(0.42)(28)(0.0752)(80.67)^2}{32.2}$$

$$= 89.4 \text{ lbf}$$

The total resisting force is

$$89.4 + (0.01)(3300) = 122.4 \text{ lbf}$$

The power consumed is

$$P = Fv$$

$$= \frac{(122.4 \text{ lbf})\left(80.67 \dfrac{\text{ft}}{\text{sec}}\right)}{778 \dfrac{\text{ft-lbf}}{\text{BTU}}}$$

$$= 12.69 \text{ BTU/sec}$$

Available power from fuel is

$$(0.28)(115{,}000) = 32{,}200 \text{ BTU/gal}$$

Fuel consumption at 55 mph is

$$\frac{12.69}{32{,}200} = 3.94 \times 10^{-4} \text{ gal/sec}$$

PROFESSIONAL PUBLICATIONS, INC. ● Belmont, CA

In 1 sec the car travels

$$\left(55\ \frac{\text{mi}}{\text{hr}}\right)\left(\frac{1\ \text{hr}}{3600\ \text{sec}}\right) = 0.0153\ \text{mi/sec}$$

Therefore, fuel consumption is

$$\frac{3.94 \times 10^{-4}\ \dfrac{\text{gal}}{\text{sec}}}{0.0153\ \dfrac{\text{mi}}{\text{sec}}} = \boxed{2.58 \times 10^{-2}\ \text{gal/mi}}$$

(This corresponds to 38.8 mpg.)

(b) At 65 mph,

$$v = 95.33\ \text{ft/sec}$$
$$D = 124.8\ \text{lbf}$$

The resisting force is

$$124.8 + 33 = 157.8\ \text{lbf}$$

$$P = \frac{(157.5)(95.33)}{778}$$

$$= 19.3\ \text{BTU/sec}$$

Fuel consumption at 65 mph is

$$\frac{19.3\ \dfrac{\text{BTU}}{\text{sec}}}{32{,}200\ \dfrac{\text{BTU}}{\text{gal}}} = 6 \times 10^{-4}\ \text{gal/sec}$$

In 1 sec the car travels

$$\frac{65}{3600} = 0.0181\ \text{mi/sec}$$

Fuel consumption is

$$\frac{6 \times 10^{-4}}{0.0181} = 3.32 \times 10^{-2}\ \text{gal/min}$$

(This corresponds to 30.1 mi/gal.)

The percent increase in fuel consumed per mile is

$$\frac{3.32 \times 10^{-2} - 2.58 \times 10^{-2}}{2.58 \times 10^{-2}} = \boxed{0.287\ (28.7\%)}$$

HYDRAULIC MACHINES

Warmups

1.
$$\left(72 \frac{\text{gal}}{\text{min}}\right)\left(2.228 \times 10^{-3} \frac{\frac{\text{ft}^3}{\text{sec}}}{\frac{\text{gal}}{\text{min}}}\right)$$

$$= \boxed{0.1604 \text{ ft}^3/\text{sec}}$$

2.
$$\text{power} = \frac{Q h_a (\text{SG})_{\text{oil}}}{3960}$$

$$Q = 37 \text{ gal/min}$$

$$h_a = \frac{\Delta p}{\gamma} = \frac{\left(40 \frac{\text{lbf}}{\text{in}^2}\right)\left(144 \frac{\text{in}^2}{\text{ft}^2}\right)}{(\text{SG})_{\text{oil}} \left(62.4 \frac{\text{lbf}}{\text{ft}^3}\right)}$$

$$= \frac{92.308}{(\text{SG})_{\text{oil}}}$$

$$\text{power} = \frac{(37)\left(\dfrac{92.308}{\text{SG}}\right)(\text{SG})}{3960}$$

$$= \boxed{0.862 \text{ hp}}$$

3.
$$(\text{power})_2 = (0.5)\left(\frac{2000}{1750}\right)^3$$

$$= \boxed{0.746 \text{ hp} \quad [\text{say } \tfrac{3}{4}\text{-hp motor}]}$$

4.
$$n = 900 \text{ rpm}$$

$$Q = \left(300 \frac{\text{gal}}{\text{sec}}\right)\left(60 \frac{\text{sec}}{\text{min}}\right)$$

$$= 18{,}000 \text{ gal/min}$$

$$h_a = 20 \text{ ft}$$

$$n_s = \frac{900\sqrt{18{,}000}}{(20)^{0.75}}$$

$$= \boxed{12{,}768}$$

5. Assume that each stage adds 150 ft of head. The suction lift is 10 ft. For a single-suction pump, from "Upper Limits of Specific Speed" chart (App. 4-B),

$$n_s \approx \boxed{2050 \text{ rpm}}$$

Concentrates

1. Assume schedule-40 pipe.

$$D_i = 0.3355 \text{ ft}$$

$$A_i = 0.08841 \text{ ft}^2$$

$$\text{v} = \frac{Q}{A} = \frac{1.25}{0.08841}$$

$$= 14.139 \text{ ft/sec}$$

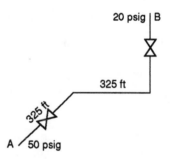

Assume regular screwed steel fittings. The approximate equivalent lengths are

fitting	length
elbows	$2 \times 13 = 26.0$ ft
gate valve	$1 \times 2.5 = 2.5$ ft
check valve	$1 \times 38 = 38.0$ ft
	66.5 ft

Assume 70°F, so $\nu = 1.059 \times 10^{-5}$.

$$\text{Re} = \frac{D\text{v}}{\nu} = \frac{(0.3355)(14.139)}{1.059 \times 10^{-5}}$$

$$= 4.479 \times 10^5$$

$\epsilon = 0.0002$ ft. So,

$$\frac{\epsilon}{D} = \frac{0.0002}{0.3355} \approx 0.0006$$

From the Moody friction factor chart, $f = 0.0185$.

$$h_f = \frac{(0.0185)(700 + 66.5)(14.139)^2}{(2)(0.3355)(32.2)}$$

$$= 131.20 \text{ ft}$$

The head added is

$$h_a = \frac{p_2}{\gamma} + \frac{\text{v}_2^2}{2g} + z_2 + h_f - \frac{p_1}{\gamma} - \frac{\text{v}_1^2}{2g} - z_1$$

But $v_1 = 0$ and $z_1 = 0$.

$$h_a = \frac{(20 + 14.7)(144)}{62.3} + \frac{(14.139)^2}{(2)(32.2)} + 50$$
$$+ 131.20 - \frac{(50 + 14.7)(144)}{62.3}$$
$$= 114.96 \text{ ft}$$

Assume 70°F water.

$$\rho = 62.3 \text{ lbm/ft}^3$$
$$\dot{m} = \left(62.3 \frac{\text{lbm}}{\text{ft}^3}\right)\left(1.25 \frac{\text{ft}^3}{\text{sec}}\right)$$
$$= 77.875 \text{ lbm/sec}$$
$$\text{hp} = \left[\frac{\left(77.875 \frac{\text{lbm}}{\text{sec}}\right)(114.96 \text{ ft})}{550 \frac{\text{ft-lbf}}{\text{hp-sec}}}\right]\left(\frac{g}{g_c}\right)$$
$$= \boxed{16.28 \text{ hp}}$$

2. The propeller is limited by cavitation. Disregarding a safety factor, this will occur when

$$h_{\text{atmosphere}} - h_{\text{velocity}} < h_{\text{vapor pressure}}$$

Assume the density of seawater is 64.0 lbm/ft³.

$$h_{\text{atmosphere}} = \frac{(14.7)(144)}{64.0} = 33.075 \text{ ft}$$
$$h_{\text{depth}} = 8$$
$$h_{\text{velocity}} = \frac{v_{\text{propeller}}^2}{2g} = \frac{(4.2v_{\text{boat}})^2}{(2)(32.2)}$$
$$= 0.2739v_{\text{boat}}^2$$

The water pressure of 70°F freshwater is found from steam tables.

$$h_{\text{vapor,fresh}} = \frac{\left(0.3631 \frac{\text{lbf}}{\text{in}^2}\right)(144)}{62.3} = 0.839 \text{ ft}$$

Raoult's law predicts the actual vapor pressure of the solution.

$$p_{\text{vapor,solution}} = (p_{\text{vapor,solvent}})(\text{mole fraction of solvent})$$

Assume $2\frac{1}{2}\%$ salt by weight. 100 lbm of seawater will yield 2.5 lbm salt and 97.5 lbm water. The molecular weight of salt is $23.0 + 35.5 = 58.5$. The number of moles of salt in 100 lbm of seawater is

$$n_s = \frac{2.5}{58.5} = 0.043$$

Similarly, the molecular weight of water is $(2)(1) + 16 = 18$. The number of moles of water is

$$n_w = \frac{97.5}{18} = 5.417$$

The mole fraction of water is

$$\frac{5.417}{5.417 + 0.043} = 0.992$$

So,
$$p_{\text{vapor,sea water}} = (0.992)(0.839)$$
$$= 0.832 \text{ ft}$$

Then,
$$8 + 33.075 - 0.2739v^2 = 0.832$$
$$v = \boxed{12.12 \text{ ft/sec}}$$

3.

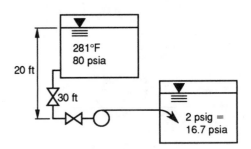

Assume schedule-40 pipe.

$$D_i = 0.1342 \text{ ft}$$
$$A_i = 0.01414 \text{ ft}^2$$
$$Q = \left(100 \frac{\text{gal}}{\text{min}}\right)(2.228 \times 10^{-3})$$
$$= 0.2228 \text{ ft}^3/\text{sec}$$
$$v = \frac{Q}{A} = \frac{0.2228}{0.01414}$$
$$= 15.76 \text{ ft/sec}$$

Assume regular screwed steel fittings. The approximate equivalent fitting lengths are

fitting		length
inlet		3.1 ft
elbows	$2 \times 3.4 =$	6.8 ft
gate valve	$2 \times 1.2 =$	2.4 ft
		12.3 ft

The total equivalent length is $30 + 12.3 = 42.3$ ft.

For steel,
$$\epsilon = 0.0002 \text{ ft}$$
$$\frac{\epsilon}{D} = \frac{0.0002}{0.1342} = 0.0015$$

At 281°F, $\nu = 0.239 \times 10^{-5}$ ft^2/sec.

$$\text{Re} = \frac{vD}{\nu} = \frac{(15.76)(0.1342)}{0.239 \times 10^{-5}} = 8.85 \times 10^5$$

From a Moody friction factor chart,

$$f = 0.022$$

$$h_f = \frac{(0.022)(42.3)(15.76)^2}{(2)(0.1342)(32.2)} = 26.74 \text{ ft}$$

At 281°F, from steam tables,

$$\rho = \frac{1}{0.01727} = 57.9 \text{ lbm/ft}^3$$

$$p_{\text{vapor}} = 50.04 \text{ in}^2$$

$$\text{NPSHA} = \frac{(80 \text{ psia})\left(144 \dfrac{\text{in}^2}{\text{ft}^2}\right)}{57.9 \dfrac{\text{lbf}}{\text{ft}^3}} + 20 - 26.74$$

$$- \frac{(50.04)(144)}{57.9}$$

$$= 67.77 \text{ ft}$$

Since NPSHR = 10 ft, the pump will not cavitate at any time as long as the pressure remains at 80 psia. Notice that the discharge line does not affect NPSHA.

4.

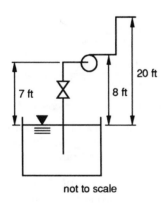

not to scale

Assume screwed steel fittings. The approximate equivalent lengths of the fittings are

fitting	length
inlet	8.5 ft
check valve	19.0 ft
elbows	$3 \times 3.6 = $ 10.8 ft
	38.3 ft

The total equivalent length of the 2-in line is

$$L_e = 12 + 38.3 + 80 = 130.3 \text{ ft}$$

Assume schedule-40 pipe.

$$D_i = 0.1723 \text{ ft}$$

$$A_i = 0.0233 \text{ ft}^2$$

Since the flow rate is unknown, it must be assumed in order to find v. Assume 90 gal/min.

$$\left(90 \frac{\text{gal}}{\text{min}}\right)\left(0.002228 \frac{\frac{\text{ft}^3}{\text{sec}}}{\frac{\text{gal}}{\text{min}}}\right) = 0.20052 \text{ ft}^3/\text{sec}$$

$$v = \frac{Q}{A} = \frac{0.20052}{0.0233} = 8.606 \text{ ft/sec}$$

$$\nu = 1.059 \times 10^{-5} \text{ ft}^2/\text{sec}$$

$$\text{Re} = \frac{Dv}{\nu} = \frac{(0.1723)(8.606)}{1.059 \times 10^{-5}} = 1.4 \times 10^5$$

$$\epsilon = 0.0002 \text{ ft}$$

$$\frac{\epsilon}{D} = \frac{0.0002}{0.1723} = 0.0012$$

From the Moody friction factor chart, $f = 0.022$. At 90 gal/min, the friction loss in the line is

$$h_f = \frac{(0.022)(130.3)(8.606)^2}{(2)(0.1723)(32.2)} = 19.1 \text{ ft}$$

At 90 gal/min, the velocity head is

$$h_v = \frac{v^2}{2g} = \frac{(8.606)^2}{(2)(32.2)} = 1.2 \text{ ft}$$

In general, the velocity head is

$$h_v = (1.2)\left(\frac{Q_2}{90}\right)^2$$

$$\frac{H_2}{H_1} = \left(\frac{Q_2}{Q_1}\right)^2$$

The total system head is

$$H = \Delta z + h_v + h_f$$

$$= 20 + (1.2 + 19.1)\left(\frac{Q_2}{90}\right)^2$$

Q_2	H	Q_2	H
0	20.0	60	29.0
10	20.3	70	32.3
20	21.0	80	36.0
30	22.3	90	40.3
40	24.0	100	45.0
50	26.3	110	50.3

The intersection point is probably not in an efficient range for the pump. A different pump should be used.

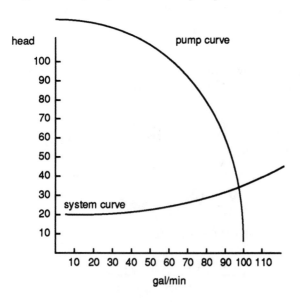

5. $\dot{V} = \dfrac{\left(8 \times 10^6 \ \dfrac{\text{gal}}{\text{day}}\right) \left(0.002228 \ \dfrac{\dfrac{\text{ft}^3}{\text{sec}}}{\dfrac{\text{gal}}{\text{min}}}\right)}{\left(24 \ \dfrac{\text{hr}}{\text{day}}\right) \left(60 \ \dfrac{\text{min}}{\text{hr}}\right)}$

$= 12.377 \ \text{ft}^3/\text{sec}$

Assume schedule-40 pipe.

pipe	D_i	A_i
8″	0.6651	0.3474
12″	0.9948	0.7773
16″	1.25	1.2272

The velocity in the inlet pipe is

$$\text{v} = \frac{Q}{A} = \frac{12.377}{0.3474} = 35.63 \ \text{ft/sec}$$

Assume 70°F, so $\nu = 1.59 \times 10^{-5}$. Then,

$$\text{Re} = \frac{D\text{v}}{\nu} = \frac{(0.6651)(35.63)}{1.059 \times 10^{-5}} = 2.24 \times 10^6$$

For steel pipe,

$$\epsilon = 0.0002$$
$$\frac{\epsilon}{D} = \frac{0.0002}{0.6651} = 0.0003$$

From the Moody fraction factor chart,

$$f = 0.015$$
$$h_{f,1} = \frac{(0.015)(1000)(35.63)^2}{(2)(0.6651)(32.2)} = 444.6 \ \text{ft}$$

For the outlet pipe,

$$\text{v} = \frac{Q}{A} = \frac{12.377}{0.7773} = 15.92 \ \text{ft/sec}$$

$$\text{Re} = \frac{(0.9948)(15.92)}{1.059 \times 10^{-5}} = 1.5 \times 10^6$$

Using $\epsilon = 0.0002$,

$$\frac{\epsilon}{D} = \frac{0.0002}{0.9948} = 0.0002$$
$$f = 0.014$$
$$h_{f,2} = \frac{(0.014)(1500)(15.92)^2}{(2)(0.9948)(32.2)} = 83.1 \ \text{ft}$$

Assume a 50% split through the two branches. In the upper branch,

$$\text{v} = \frac{Q}{A} = \frac{\left(\frac{1}{2}\right)(12.377)}{0.3474} = 17.81 \ \text{ft/sec}$$
$$\text{Re} = \frac{(0.6651)(17.81)}{1.059 \times 10^5} = 1.1 \times 10^6$$

Using $\epsilon/D = 0.0003$, $f = 0.015$.

For the 16-in pipe,

$$\text{v} = \frac{Q}{A} = \frac{\left(\frac{1}{2}\right)(12.377)}{1.2272}$$
$$= \frac{6.189}{1.2272} = 5.04$$

$\epsilon = 0.0002$, so

$$\frac{\epsilon}{D} = \frac{0.0002}{1.25} = 0.00016$$
$$\text{Re} = \frac{(1.25)(5.04)}{1.059 \times 10^{-5}} = 5.9 \times 10^5$$

From the Moody friction factor chart, $f \approx 0.015$.

These values of f for the two branches are fairly insensitive to changes in Q, so they will be used for the rest of the problem in the upper branch. Assuming 4 MGD,

$$h_{f,\text{upper}} = \frac{(0.015)(500)(17.81)^2}{(2)(0.6651)(32.2)} = 55.5$$

The loss for any other flow is

$$H_2 = H_1 \left(\frac{Q_2}{Q_1}\right)^2$$
$$= (55.5) \left(\frac{Q}{6.189}\right)^2 = 1.45 Q^2$$

Similarly, for the lower branch, in the 8-in section,

$$h_{f,\text{lower,8 in}} = \frac{(0.015)(250)(17.81)^2}{(2)(0.6651)(32.2)}$$
$$= 27.8 \ \text{ft}$$

In the 16-in section,

$$h_{f,\text{lower},16\text{ in}} = \frac{(0.015)(1000)(5.04)^2}{(2)(1.25)(32.2)}$$

$$= 4.7 \text{ ft}$$

The total loss in the lower branch is

$$h_{f,\text{lower}} = 27.8 + 4.7 = 32.5$$

For any other flow, the loss will be

$$H_2 = (32.5)\left(\frac{Q}{6.189}\right)^2 = (0.85)(Q)^2$$

Let x be the fraction flowing in the upper branch. Then, because the friction losses are equal,

$$(1.45)[(x)(12.377)]^2 = (0.85)[(1-x)(12.377)]^2$$

$$x = 0.432$$

Then,

$$Q_{\text{upper}} = (0.432)(12.377) = \boxed{5.347 \text{ ft}^3/\text{sec}}$$

$$Q_{\text{lower}} = (1 - 0.432)(12.377) = \boxed{7.03 \text{ ft}^3/\text{sec}}$$

$$H_2 = (1.45)(5.347)^2 = 41.5 \text{ ft}$$

$$h_{f,\text{total}} = 444.6 + 41.5 + 83.1$$

$$= \boxed{569.2}$$

6. (a) At 70°F, the head dropped is

$$H = \frac{\left(500\ \dfrac{\text{lbf}}{\text{in}^2} - 30\ \dfrac{\text{lbf}}{\text{in}^2}\right)\left(144\ \dfrac{\text{in}^2}{\text{ft}}\right)}{62.4\ \dfrac{\text{lbf}}{\text{ft}^3}} = 1084.6 \text{ ft}$$

The specific speed of a turbine is

$$n_s = \frac{n\sqrt{\text{bhp}}}{H^{1.25}} = \frac{1750\sqrt{250}}{(1084.6)^{1.25}}$$

$$= 4.446 \text{ rpm}$$

> Since the lowest suggested value of n_s for a reaction turbine is 10, recommend an impulse (Pelton) wheel.

(b) $$Q = Av = \left(\frac{\pi}{4}\right)\left(\frac{4}{12}\right)^2 (35)$$

$$= 3.054 \text{ ft}^3/\text{sec}$$

$$Q' = \frac{(35-10)(3.054)}{35} = 2.181 \text{ ft}^3/\text{sec}$$

Assuming $\gamma = 62.4$ lbf/ft^3,

$$F_x = \left[\frac{(2.181)(62.4)}{32.2}\right](35-10)(\cos 80° - 1)$$

$$= -87.32 \text{ lbf}$$

$$F_y = \left[\frac{(2.181)(62.4)}{32.2}\right](35-10)(\sin 80°)$$

$$= 104.06 \text{ lbf}$$

$$R = \sqrt{(-87.32)^2 + (104.06)^2} = \boxed{135.84 \text{ lbf}}$$

7.

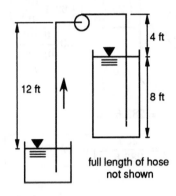

full length of hose
not shown

$$80 \text{ gal/min} = (80)(0.002228)$$

$$= 0.1782 \text{ ft}^3/\text{sec}$$

$$A = \left(\frac{\pi}{4}\right)\left(\frac{2}{12}\right)^2 = 0.0218$$

$$v = \frac{Q}{A} = \frac{0.1782}{0.0218}$$

$$= 8.174 \text{ ft/sec}$$

Use 80°F water data.

$$\nu = 0.93 \times 10^{-5}$$

$$\text{Re} = \frac{vD}{\nu} = \frac{(8.174)\left(\dfrac{2}{12}\right)}{0.93 \times 10^{-5}}$$

$$= 1.46 \times 10^5$$

Assume that the rubber hose is smooth. From the Moody friction factor chart,

$$f = 0.016$$

$$h_f = \frac{(0.016)(50)(8.174)^2}{(2)\left(\dfrac{2}{12}\right)(32.2)} = 5 \text{ ft}$$

$$h_v = \frac{(8.174)^2}{(2)(32.2)} = 1.0$$

$$h_a = 5 + 1 + 12 - 4$$

$$= \boxed{14 \text{ ft}}$$

This neglects entrance losses. (Note that the hose is 50 ft long even though the lift is only 12 ft.)

Timed

1. (a) The total entering head is

$$H = 92.5 + \frac{(12)^2}{(2)(32.2)} + 5.26 = \boxed{100 \text{ ft}}$$

(b) The water horsepower is

$$\left(\frac{\dot{m}H}{550}\right)\left(\frac{g}{g_c}\right) = \left[\frac{\left(25\frac{\text{ft}^3}{\text{sec}}\right)\left(62.4\frac{\text{lbm}}{\text{ft}^3}\right)(100\text{ ft})}{550\frac{\text{ft-lbf}}{\text{hp-sec}}}\right]$$

$$\times \left(\frac{g}{g_c}\right)$$

$$= 283.6 \text{ hp}$$

$$\eta = \frac{250}{283.6} = \boxed{0.882 \ (88.2\%)}$$

(c)

$$n_2 = n_1\sqrt{\frac{H_2}{H_1}} = 610\sqrt{\frac{225}{100}}$$

$$= \boxed{915 \text{ rpm}}$$

(d)

$$(\text{power})_2 = (\text{power})_1\left(\frac{H_2}{H_1}\right)^{1.5}$$

$$= (250)\left(\frac{225}{100}\right)^{1.5}$$

$$= \boxed{843.8 \text{ hp}}$$

(e)

$$Q_2 = Q_1\sqrt{\frac{H_2}{H_1}} = 25\sqrt{\frac{225}{100}}$$

$$= \boxed{37.5 \text{ ft}^3/\text{sec}}$$

2. (a) For a 1,000,000 BTU/hr system, the enthalpy drop across the radiator is

$$167.99 - 147.92 = 20.07 \text{ BTU/lbm}$$

The mass flow rate through the system is

$$\frac{1,000,000 \frac{\text{BTU}}{\text{hr}}}{\left(60\frac{\text{min}}{\text{hr}}\right)\left(20.07\frac{\text{BTU}}{\text{lbm}}\right)} = 830.4 \text{ lbm/min}$$

At the bulk temperature of 190°F,

$$\rho = \frac{1}{0.01657} = 60.35 \text{ lbm/ft}^3$$

The volume flow rate is

$$Q = \frac{\left(830.4\frac{\text{lbm}}{\text{min}}\right)\left(7.48\frac{\text{gal}}{\text{ft}^3}\right)}{\left(60.35\frac{\text{lbm}}{\text{ft}^3}\right)} = 102.9 \text{ gal/min}$$

Plot the pump curves and determine the head losses at 102.9 gal/min.

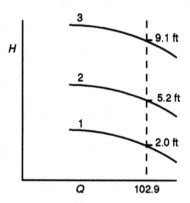

The loss per foot for the three pumps is

$$\text{pump 1:} \quad \frac{(2.0)(12)}{420} = 0.057 \text{ in/ft} \quad \text{[too low]}$$

$$\text{pump 2:} \quad \frac{(5.2)(12)}{420} = 0.15 \text{ in/ft} \quad \text{[too low]}$$

$$\text{pump 3:} \quad \frac{(9.1)(12)}{420} = 0.26 \text{ in/ft} \quad \text{[ok]}$$

$$\boxed{\text{Choose pump 3.}}$$

Since the pipe material was not specified, assume $f = 0.20$ (good for turbulent flow in steel pipe 1-in to 3-in in diameter).

Keep the velocity between 2–4 ft/sec for noise.

$$h_f = \frac{fLv^2}{2Dg}$$

$$D = \frac{fLv^2}{2h_fg} = \frac{(0.02)(420)(4)^2}{(2)(9.1)(32.2)}$$

$$= 0.229 \text{ ft} \ (2.75 \text{ in})$$

$$\boxed{\text{Say } D = 3 \text{ in.}}$$

Try a different approach.

$$A = \frac{Q}{v} = \frac{102.9 \ \frac{\text{gal}}{\text{min}}}{\left(60 \ \frac{\text{sec}}{\text{min}}\right)\left(7.48 \ \frac{\text{gal}}{\text{ft}^3}\right)\left(4 \ \frac{\text{ft}}{\text{sec}}\right)}$$

$$= 0.057 \ \text{ft}^2$$

$$D = \sqrt{\frac{4A}{\pi}} = \sqrt{\frac{(4)(0.057)}{\pi}}$$

$$= 0.27 \ \text{ft} \quad (3.23 \ \text{in})$$

Say $D = 3\frac{1}{2}$ in (size may not be available).

The actual pipe inside diameter will depend on whether copper, steel, or cast iron pipe is used.

(b) For a 300,000 BTU/hr system, proceed similarly.

$$Q = \frac{(300{,}000)(7.48)}{(60)(20.07)(60.35)} = 30.88 \ \text{gal/min}$$

The head loss at 30.88 gal/min for each pump is

pump 1: $\dfrac{(5.5)(12)}{420} = 0.157 \ \text{in/ft}$ [too low]

pump 2: $\dfrac{(9.0)(12)}{420} = 0.257 \ \text{in/ft}$ [ok]

pump 3: $\dfrac{(13.4)(12)}{420} = 0.38 \ \text{in/ft}$ [ok]

Pumps 2 and 3 meet the specifications.

$$A = \frac{30.88}{(60)(7.48)(4)} = 0.0172 \ \text{ft}^2$$

$$D = \sqrt{\frac{(4)(0.0172)}{\pi}} = 0.148 \ \text{ft} \quad (1.78 \ \text{in})$$

Say $1\frac{3}{4}$-in pipe (size may not be available).

3. (a) The equivalent lengths are assumed to be in feet of the sections' diameters. Since the diameters are different, the equivalent lengths cannot be added together. It is necessary to convert the lengths to pressure drops. Although the Darcy equation could be used, it is more convenient to use standard tables of friction losses for the type-L tubing.

circuit	loss per ft	equivalent length	total loss
5-1-P-2	0.3 in wg	40	12 in
2-4	0.47	70	32.9
2-3	0.40	55	22
3-4	0.47	65	30.6
3-5	0.47	60	28.2
4-5	0.40	50	20

These head losses are for cold water and are greater than hot water losses. The required horsepower will be less for hot water. However, the system may be tested with cold water, and the horsepower must be adequate for this.

Redraw the pipe network.

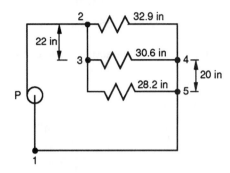

The pump must be able to handle the longest run.

run	
2-4-5:	$32.9 + 20 = 52.9$
2-3-4-5:	$22 + 30.6 + 20 = 72.6$
2-3-5:	$22 + 28.2 = 50.2$

The longest run is 2-3-4-5 with a loss of 72.6 in. The total circuit loss is

$$h_t = \frac{72.6 + 12}{12} = 7.05 \ \text{ft}$$

Note that the loss between points 2 and 5 must be the same regardless of the path. Therefore, flow restriction valves must exist in the other two runs in order to match the longest run's loss.

The water horsepower is

$$\text{whp} = \frac{(7.05 \ \text{ft})\left(60 \ \frac{\text{gal}}{\text{min}}\right)}{3960} = 0.107 \ \text{hp}$$

$$\text{bhp} = \frac{\text{whp}}{\eta} = \frac{0.107}{0.45}$$

$$= 0.238 \ \text{hp}$$

(b) Use a $\frac{1}{4}$-hp or larger motor.

(c) $\dot{m} = \left(60 \ \dfrac{\text{gal}}{\text{min}}\right)\left(0.1337 \ \dfrac{\text{ft}^3}{\text{gal}}\right)\left(60.1 \ \dfrac{\text{lbm}}{\text{ft}^3}\right)$

$= 482 \ \text{lbm/min}$

$q = \dot{m} c_p \Delta T$

$= \left(482 \ \dfrac{\text{lbm}}{\text{min}}\right)\left(60 \ \dfrac{\text{min}}{\text{hr}}\right)\left(1 \ \dfrac{\text{BTU}}{\text{lbm-}^\circ\text{F}}\right)(20^\circ\text{F})$

$= 578{,}400 \ \text{BTU/hr}$

4. Assume regular 90° elbows. The equivalent lengths of pipe and fittings are

$$500 + (6)(4.4) + (2)(2.8) = 532 \text{ ft}$$

For 3-in steel schedule-40 pipe,

$$A_i = 0.05134 \text{ ft}^3$$
$$D_i = 0.2557 \text{ ft}$$

Assume 100 gal/min. (The actual value chosen for this first estimate is not important.)

$$v = \frac{Q}{A}$$
$$= \frac{\left(100 \ \frac{\text{gal}}{\text{min}}\right)\left(0.1337 \ \frac{\text{ft}^3}{\text{gal}}\right)}{(0.05134 \text{ ft}^2)\left(60 \ \frac{\text{sec}}{\text{min}}\right)}$$
$$= 4.34 \text{ ft/sec}$$
$$\text{Re} = \frac{Dv}{\nu} = \frac{(0.2557)(4.34)}{6 \times 10^{-6}}$$
$$= 1.85 \times 10^5$$

Use $\epsilon = 0.0002$ ft.

$$\frac{\epsilon}{D} = \frac{0.0002}{0.2557} = 0.0008$$

From the Moody chart, $f \approx 0.02$ (goes to 0.018 at full turbulence, which this will be at higher flows). Use $f = 0.018$.

$$h_f = \frac{(0.018)(532)(4.34)^2}{(2)(0.2557)(32.2)}$$
$$= 10.95 \text{ ft of gasoline}$$

This neglects the small velocity head, although it can be included. The other system curve points can be found using

$$\frac{h_1}{h_2} = \left(\frac{Q_1}{Q_2}\right)^2$$
$$h_2 = (10.95)\left(\frac{Q_2}{100}\right)^2 = 0.001095 Q_2^2$$

Q	h_f	$h+60$
100	10.95	71.0
200	43.8	103.8
300	98.6	158.6
400	175.2	235.2
500	273.8	333.8
600	394.2	454.2

The head does not depend on the fluid being pumped. Therefore, the water curve can be used.

(a) Plotting the system and pump curves gives

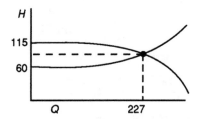

$$\boxed{\begin{array}{l} Q = 227 \text{ gal/min} \\ h = 115 \text{ ft} \end{array}}$$

(b) The cost per hour is

$$\left(0.045 \ \frac{\$}{\text{kW-hr}}\right)$$
$$\times \left[\frac{\left(227 \ \frac{\text{gal}}{\text{min}}\right)\left(0.1337 \ \frac{\text{ft}^3}{\text{gal}}\right)\left(62.4 \ \frac{\text{lbf}}{\text{ft}^3}\right)(0.7)}{\left(33{,}000 \ \frac{\text{ft-lbf}}{\text{hp-min}}\right)(0.88)(0.88)}\right.$$
$$\left. \times (115 \text{ ft})\left(0.7457 \ \frac{\text{kW}}{\text{hp}}\right)(1 \text{ hr})\right]$$
$$= \boxed{\$0.20}$$

5.

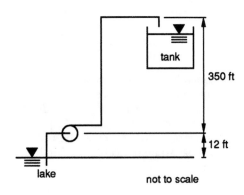

$$Q = 10{,}000 \text{ gal/hr}$$
$$= 1337 \text{ ft}^3/\text{hr}$$
$$\eta = 0.7$$

Assume 60°F and 14.7 psia.

For 4-in schedule-40 steel pipe,

$$D_i = 0.3355 \text{ ft}$$
$$A = 0.08841 \text{ ft}^2$$
$$v = \frac{Q}{A} = \frac{\left(1337 \frac{\text{ft}^3}{\text{hr}}\right)\left(\frac{1 \text{ hr}}{3600 \text{ sec}}\right)}{0.08841 \text{ ft}^2}$$
$$= 4.2 \text{ ft/sec}$$

For 60°F water,

$$\nu = 1.217 \times 10^{-5} \text{ ft}^2/\text{sec}$$
$$\text{Re} = \frac{D_e v}{\nu} = \frac{(0.3355)(4.2)}{1.217 \times 10^{-5}}$$
$$= 1.16 \times 10^5$$

For welded and seamless steel,

$$\epsilon \approx 0.0002 \text{ ft}$$
$$\frac{\epsilon}{D} = \frac{0.0002}{0.3355} = 5.96 \times 10^{-4}$$
$$\approx 0.0006$$

Using the Moody friction factor chart,

$$f = 0.02$$
$$h_f = \frac{f L_e v^2}{2Dg}$$
$$= \frac{(0.02)(7000 \text{ ft})\left(4.2 \frac{\text{ft}}{\text{sec}}\right)^2}{(2)(0.3355 \text{ ft})\left(32.2 \frac{\text{ft}}{\text{sec}^2}\right)}$$
$$= 114.3 \text{ ft}$$
$$h_v = \frac{v^2}{2g} = \frac{(4.2)^2}{(2)(32.2)}$$
$$= 0.27 \text{ ft}$$

The head added by the pump is

$$h_a = 12 + 350 + 0.27 + 114.3 = 476.6 \text{ ft}$$
$$\text{whp} = \frac{h_a Q (\text{SG})}{3960}$$
$$= \frac{(476.6 \text{ ft})\left(10,000 \frac{\text{gal}}{\text{hr}}\right)\left(\frac{1 \text{ hr}}{60 \text{ min}}\right)(1)}{3960}$$
$$= 20.06 \text{ hp}$$
$$\text{ehp} = \frac{\text{whp}}{\eta} = \frac{20.06}{0.7}$$
$$= 28.7 \text{ hp}$$
$$\text{power} = (0.7457)(28.7) = 21.4 \text{ kW}$$

(a) At \$0.04/kW-hr, power costs

$$(21.4)(\$0.04) = \boxed{\$0.856 \text{ per hour}}$$

(b) motor hp $\approx$ ehp = 28.7 hp

$$\approx \boxed{\text{30-hp motor required}}$$

(c) $\text{NPSHA} = h_a + h_s - h_{f(s)} - h_{\text{vp}}$

Using $\gamma = 62.37 \text{ lbf/ft}^3$,

$$h_a = \frac{p_a}{\gamma} = \frac{(14.7)(144)}{62.37}$$
$$= 33.94 \text{ ft}$$
$$h_f = \left(\frac{300}{7000}\right) h_{f,\text{total}}$$
$$= \left(\frac{300}{7000}\right)(114.3)$$
$$= 4.9 \text{ ft}$$

From steam tables,

$$h_{\text{vp}} = 0.59 \text{ ft}$$
$$\text{NPSHA} = 33.94 - 12 - 4.9 - 0.59$$
$$= \boxed{16.45 \text{ ft}}$$

6.

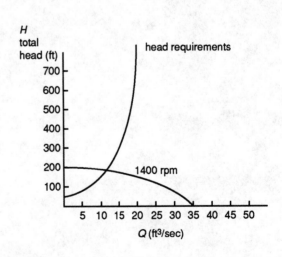

$$H = 30 + 2Q^2 \quad \text{[head requirements]}$$

(a) From the intersection of the graphs, at 1400 rpm the flow rate is approximately 9 ft^3/sec.

$$Q = \left(9 \ \frac{ft^3}{sec}\right)\left(448.8 \ \frac{\frac{gal}{min}}{\frac{ft^3}{sec}}\right)$$

$$= \boxed{4039 \ gal/min}$$

$$h_a = 30 + (2)(9)^2 = 192 \ ft$$

(b)
$$whp = \frac{(192)(9)(1)}{8.814} = 196 \ hp$$

$$n_s = \frac{n\sqrt{Q}}{(h_a)^{0.75}} = \frac{1400\sqrt{4039}}{(192)^{0.75}}$$

$$= 1725$$

From Fig. 4.13, $\eta \approx 83\%$. The minimum pump motor power should be

$$\frac{196 \ hp}{0.83} = \boxed{236 \ hp \ (use \ 250 \ hp)}$$

(c) Method 1: If the pump is turned at 1200 rpm,

$$Q_2 = \frac{n_2 Q_1}{n_1} = \frac{(1200)(4039)}{1400}$$

$$= \boxed{3462 \ gal/min}$$

Method 2: Use the affinity laws to modify the pump curve for operation at 1200 rpm. For any specific head loss, h_1,

$$Q_2 = Q_1\left(\frac{n_2}{n_1}\right) = Q\left(\frac{1200 \ rpm}{1400 \ rpm}\right)$$

$$= 0.86 Q_1$$

So, the pump curve is scaled down to approximately 86% of its original values.

The system curve is unchanged.

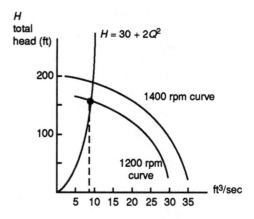

Within the ability to read the graph,

$$Q_{2, 1200 \ rpm} \approx 8 \ ft^3/sec \ (3600 \ gpm)$$

FANS AND DUCTWORK

Warmups

1. With $R = 4$, the short side is

$$a = \frac{(18 \text{ in})(4+1)^{0.25}}{(1.3)(4)^{0.625}} = \boxed{8.705 \text{ in}}$$

The long side is

$$b = Ra = (4)(8.705) = \boxed{34.82 \text{ in}}$$

2. Disregarding terminal pressure, other points on the system curve are found from

$$\frac{p_2}{p_1} = \left(\frac{Q_2}{Q_1}\right)^2$$

$$Q_2 = Q_1\sqrt{\frac{p_2}{p_1}}$$

$$= \frac{10,000\sqrt{p_2}}{\sqrt{4}}$$

$$= 5000\sqrt{p_2}$$

p_2 (in)	Q (ft^3/min)
1	5000
2	7070
3	8660
4	10,000
5	11,180
6	12,250

3. At standard conditions, $\rho = 0.075 \text{ lbm/ft}^3$. Using standard atmospheric data,

$$\rho_{5000} = \frac{p}{RT} = \frac{(12.225)(144)}{(53.3)(500.9)}$$

$$= 0.06594 \text{ lbm/ft}^3$$

The air horsepower is

$$(\text{AHP})_1 = \frac{\left(40 \dfrac{\text{ft}^3}{\text{min}}\right)(0.5 \text{ in wg})}{6356 \dfrac{\text{cm-wg}}{\text{hp}}}$$

$$= 3.147 \times 10^{-3} \text{ hp}$$

$$(\text{AHP})_2 = (3.147 \times 10^{-3})\left(\frac{1}{\frac{1}{8}}\right)^5 \left[\frac{\left(\frac{1}{2}\right)(300)}{300}\right]^3$$

$$\times \left(\frac{0.06594}{0.075}\right)$$

$$= \boxed{11.3 \text{ hp}}$$

Note that such a large extrapolation is not recommended because actual efficiency will probably be much lower.

4. From the friction/flow quantity chart, the friction loss is 0.047 in water per 100 ft, and v = 2700 ft/min. The friction loss due to the duct is

$$h_{f,1} = (0.47 \text{ in wg})\left(\frac{750}{100}\right)$$

$$= 3.525 \text{ in wg}$$

The equivalent length of each elbow is $12D$. For the two,

$$L_e = (4)\left[(12)\left(\frac{20 \text{ in}}{12 \frac{\text{in}}{\text{ft}}}\right)\right] = 80 \text{ ft}$$

This creates a friction loss of

$$h_{f,2} = (0.47)\left(\frac{80}{100}\right) = 0.376 \text{ in wg}$$

$$\frac{E}{D} = \frac{2}{20} = 0.10$$

$$C_1 = 0.2$$

$$h_{f,3} = (2)\left[(0.2)\left(\frac{2700}{4005}\right)^2\right]$$

$$= 0.182 \text{ in wg}$$

The total loss is

$$\sum h_f = 3.525 + 0.376 + 0.182$$

$$= \boxed{4.083 \text{ in wg}}$$

5. In the 18-in duct,

$$A = \frac{\left(\frac{\pi}{4}\right)(18)^2}{144} = 1.767 \text{ ft}^2$$

$$v_1 = \frac{Q}{A} = \frac{1500}{1.767}$$

$$= 848.9 \text{ ft/min}$$

In the 14-in duct,

$$A = \frac{\left(\frac{\pi}{4}\right)(14)^2}{144} = 1.069 \text{ ft}^2$$

$$v_2 = \frac{Q}{A} = \frac{1500 - 400}{1.069}$$

$$= 1029 \text{ ft/min}$$

Since $v_2 > v_1$, there is no regain. Use $R = 1.1$ to account for a 10% friction loss.

$$\Delta p_s = (1.1) \left[\frac{(848.9)^2 - (1029)^2}{(4005)^2} \right]$$

$$= \boxed{-0.023 \text{ in wg loss}}$$

Concentrates

1. Select 1600 ft/min as the main duct velocity. With 1600 ft/min and 1500 ft^3/min,

$$h_f = 0.27 \text{ in wg per 100 ft}$$
$$D_{\text{fan-A}} = 13 \text{ in}$$

The equivalent rectangular duct sides are

$$a = \frac{D(1.5+1)^{0.25}}{(1.3)(1.5)^{0.625}} = 0.75D$$
$$b = Ra = (1.5)(0.75D) = 1.125D$$
$$a_{\text{fan-A}} = (0.75)(13) = 9.75 \text{ in}$$
$$b_{\text{fan-A}} = (1.125)(13) = 14.63 \text{ in}$$

Proceeding similarly for all sections, the following table is obtained.

section	Q	D	ab^*
fan–A	1500	13	(9.8)(14.6)
A–B	1200	11.8	(8.9)(13.3)
B–C	900	11.0	(8.3)(12.4)
C–D	600	9.1	(6.8)(10.2)
D–E	400	8.0	(6)(9)
E–F	200	6.0	(4.5)(6.8)

*Actual values would be rounded to whole inches.

Assume $r/D = 1.5$ for the bends. Then, $L_e = 12D$. For the two elbows,

$$L_e = (12)\left(\frac{13}{12} + \frac{9.1}{12}\right) = 22.1 \text{ ft}$$

The distance from the fan to F is

$$15 + 45 + 30 + 30 + 20 + 10 + 20 + 20 = 190$$

Then,

$$h_f = (190 + 22.1)\left(\frac{0.27 \text{ in wg}}{100 \text{ ft}}\right)$$
$$= 0.57 \text{ in wg}$$

The fan must also supply the terminal pressure.

$$0.57 + 0.25 = \boxed{0.82 \text{ in wg}}$$

The equal friction method generally ignores velocity head contribution to static pressure. If it is included, add

$$\left(\frac{1600}{4005}\right)^2 = 0.16 \text{ in wg}$$

2. Choose 1600 ft/min as the main duct velocity.

$$Q_{\text{fan}} = (12)(300) = 3600 \text{ ft}^3/\text{min}$$

From the friction/flow quality chart with $Q = 3600$ ft^3/min and $v = 1600$ ft/min,

$$h = 0.16 \text{ in wg}/100 \text{ ft}$$
$$D_{\text{fan-1st}} = 20 \text{ in}$$

Proceeding according to the equal friction method results in the following table.

section	Q	D	section	Q	D
fan–1st	3600	20	2nd–E	1200	13.2
1st–2nd	2400	17.2	E–F	900	12
2nd–3rd	1200	13.2	F–G	600	10.1
1st–A	1200	13.2	G–H	300	7.9
A–B	900	12	I–J	900	12
B–C	600	10.1	J–K	600	10.1
C–D	300	7.9	K–L	300	7.9

The longest run is fan–L. Its length is

$$L_{\text{longest}} = 25 + 35 + 20 + 20 + 10 + 20 + 20 + 20$$
$$= 170$$

The elbows have $L_e = 14.5D$ (interpolated). So,

$$L_e = 170 + (14.5)\left(\frac{20 \text{ in} + 13.2 \text{ in}}{12 \frac{\text{in}}{\text{ft}}}\right)$$
$$= 210.1 \text{ ft}$$

The fan supplies

$$(210.1)\left(\frac{0.16 \text{ in wg}}{100 \text{ ft}}\right) + 0.15 = \boxed{0.486 \text{ in}}$$

3. Choose 1600 ft/min as the main duct velocity. Then, $Q = 3600$ ft^3/min, and from a friction/flow quantity chart,

$$h_f = \frac{16 \text{ in wg}}{100 \text{ ft}}$$
$$D = 20 \text{ in}$$

The bends have $L_e = 14.5D$. Then, the loss up to the first take-off is

$$h_{f,\text{fan--1st}} = \frac{(0.16)\left[25 + 35 + (14.5)\left(\frac{20}{12}\right)\right]}{100}$$

$$= 0.135 \text{ in wg}$$

After the first take-off,

$$Q = 3600 - 1200 = 2400 \text{ ft}^3/\text{min}$$

$$L_e = 20 \text{ ft}$$

$$\frac{L_e}{Q^{0.61}} = \frac{20}{(2400)^{0.61}}$$

$$= 0.173$$

From the static regain chart, $v_2 = 1390$. Proceeding similarly, the following table is developed.

section	Q	L_e	$\dfrac{L_e}{Q^{0.61}}$	v^{**}	D
fan--1st	3600	80.8	0.547	1600	20
1st--2nd	2400	20	0.173	1390	(18)(17.8)
2nd--I*	1200	49.3	0.65	960	(15)(15.1)
I--J	900	20	0.31	800	(15)(14.4)
J--K	600	20	0.40	650	13
K--L	300	20	0.62	500	(11)(10.5)

*L_e is found assuming $D = 16$.

$$L_e = 20 + (14.5)\left(\frac{16}{12}\right) + 10 = 49.3 \text{ ft}$$

** Solved graphically.

The fan must supply

$$0.135 \text{ in} + 0.15 \text{ in} = \boxed{0.285 \text{ in wg}}$$

Timed

1. No damper is needed in duct A. The area of section A is

$$\left(\frac{\pi}{4}\right)\left(\frac{12}{12}\right)^2 = 0.7854 \text{ ft}^2$$

The velocity in section A is

$$v_A = \frac{Q}{A} = \frac{3000 \dfrac{\text{ft}^3}{\text{min}}}{\left(60 \dfrac{\text{sec}}{\text{min}}\right)(0.7854 \text{ ft}^2)}$$

$$= 63.66 \text{ ft/sec}$$

$$= 3819.7 \text{ ft/min}$$

For four-piece ells with $r/d = 1.5$,

$$L_e \approx 14D$$

Assume $D = 10$ in. The total equivalent length of run C is

$$L_e = 50 + 10 + 10 + (2)\left[(14)\left(\frac{10 \text{ in}}{12 \frac{\text{in}}{\text{ft}}}\right)\right]$$

$$= 93.3 \text{ ft}$$

For any diameter, D, in inches of section C, the velocity will be

$$v_C = \frac{Q}{A} = \frac{2000 \dfrac{\text{ft}^3}{\text{min}}}{\left(\dfrac{\pi}{4}\right)\left(\dfrac{D}{12}\right)^2\left(60 \dfrac{\text{sec}}{\text{min}}\right)}$$

$$= \frac{6111.5}{D^2}$$

The friction loss in section C will be

$$h_{f,C} = \frac{(0.02)(93.3)\left(\dfrac{6111.5}{D^2}\right)^2}{(2)\left(\dfrac{D}{12}\right)(32.2)}$$

$$= 1.3 \times 10^7/D^5 \text{ ft of air}$$

The regain between A and C will be

$$h_{\text{regain}} = (0.65)\left(\frac{v_A^2 - v_C^2}{2g}\right)$$

$$= (0.65)\left[\frac{(63.66)^2 - \left(\dfrac{6111.5}{D^2}\right)^2}{(2)(32.2)}\right]$$

$$= 40.9 - \frac{3.77 \times 10^5}{D^4} \text{ ft of air}$$

The principle of static regain is that

$$h_f = h_{\text{regain}}$$

$$\frac{1.3 \times 10^7}{D^5} = 40.9 - \frac{3.77 \times 10^5}{D^4}$$

By trial and error, $D \approx 13.5$ in. Since $D = 10$ in was assumed to find the equivalent length of the ells, this process should be repeated.

$$L_e = 70 + (2)\left[(14)\left(\frac{13.5}{12}\right)\right]$$

$$= 101.5$$

$$h_f = \frac{(0.02)(101.5)\left(\dfrac{6111.5}{D^2}\right)^2}{(2)\left(\dfrac{D}{12}\right)(32.2)}$$

$$= \frac{1.41 \times 10^7}{D^5}$$

$$\frac{1.41 \times 10^7}{D^5} = 40.9 - \frac{3.77 \times 10^5}{D^4}$$

By trial and error, $D = 13.63$ in (say 14 in). This results in a friction loss of

$$h_f = \frac{1.41 \times 10^7}{(14)^5} = 26.2 \text{ ft of air}$$

Because the regain cancels this, the pressure loss from A to C is zero. No dampers are needed in duct C.

For any diameter, D, in inches in section B, the velocity will be

$$v_B = \frac{Q}{A} = \frac{1000}{\left(\dfrac{\pi}{4}\right)\left(\dfrac{D}{12}\right)^2 (60)}$$

$$= \frac{3055.8}{D^2}$$

The friction loss in section B will be

$$h_f = \frac{(0.02)(10)\left(\dfrac{3055.8}{D^2}\right)^2}{(2)\left(\dfrac{D}{12}\right)(32.2)}$$

$$= 3.48 \times 10^5 / D^5 \text{ ft of air}$$

The 45° take-off will also create a loss.

$$h_f = c_b \left(\frac{v_A}{4005}\right)^2 \text{ in of water}$$

At this point, assume $v_B/v_A = 1.00$. Then, from a table of minor losses, for a 45° fitting, $c_b = 0.5$.

$$h_f = (0.5)\left(\frac{3819.7}{4005}\right)^2 = 0.455 \text{ in of water}$$

$$= 31.5 \text{ ft of air}$$

The regain between A and B will be

$$h_{\text{regain}} = (0.65)\left(\frac{v_A^2 - v_B^2}{2g}\right)$$

$$= (0.65)\left[\frac{(63.66)^2 - \left(\dfrac{3055.8}{D^2}\right)^2}{(2)(32.2)}\right]$$

$$= 40.9 - \frac{9.42 \times 10^4}{D^4}$$

Set the regain equal to the loss.

$$40.9 - \frac{9.42 \times 10^4}{D^4} = \frac{3.48 \times 10^5}{D^5} + 31.5$$

By trial and error,

$$D_B = 10.77 \quad [\text{say 11 in}]$$

$$v_B = \frac{3055.8}{(11)^2} = 25.25 \text{ ft/sec}$$

$$\frac{v_B}{v_A} = \frac{25.25}{63.6}$$

$$\approx 0.4$$

Therefore, the value of c_b is still ≈ 0.5. For section B, the friction cancelled by the regain is

$$h_f = 31.5 + \frac{3.45 \times 10^5}{(11)^5} = \boxed{33.6 \text{ ft}}$$

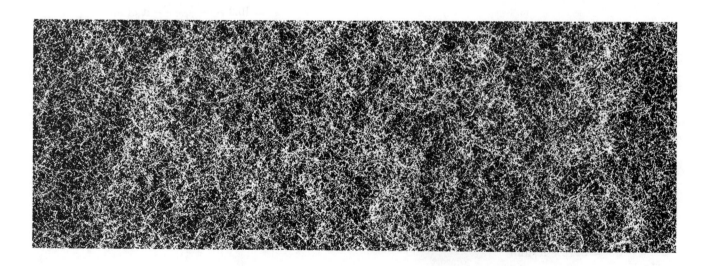

PROFESSIONAL PUBLICATIONS, INC. ● Belmont, CA

THERMODYNAMICS

Warmups

1.
$$h = 218.48 + (0.92)(945.5)$$
$$= 1088.34 \text{ BTU/lbm}$$

The molecular weight of water is 18, so

$$H = (18)(1088.34)$$
$$= \boxed{19{,}590 \text{ BTU/lbmole}}$$

2.
$$h_1 = 393.84 + (0.95)(809)$$
$$= 1162.4 \text{ BTU/lbm}$$

From the Mollier diagram for an isentropic process,

$$h_2 = 1031 \text{ BTU/lbm}$$

The maximum work output is

$$h_1 - h_2 = 1162.4 - 1031$$
$$= \boxed{131.4 \text{ BTU/lbm}}$$

3. The average value of c_p for iron is approximately

$$c_p = 0.10$$
$$Q = \left(0.10 \ \frac{\text{BTU}}{\text{lbm-°F}}\right)(780°\text{F} - 80°\text{F})$$
$$= \boxed{70.0 \text{ BTU/lbm}}$$

4.
$$c_p = 0.250 \text{ BTU/lbm-°F}$$
$$R = 53.3 \text{ ft-lbf/lbm-°R}$$
$$c_v = c_p - \frac{R}{J}$$
$$= 0.250 \ \frac{\text{BTU}}{\text{lbm-°F}} - \frac{53.3 \ \dfrac{\text{ft-lbf}}{\text{lbm-°R}}}{778 \ \dfrac{\text{ft-lbf}}{\text{BTU}}}$$
$$= 0.1815 \text{ BTU/lbm-°F}$$
$$k = \frac{c_p}{c_v} = \frac{0.250}{0.1815}$$
$$= \boxed{1.377}$$

5.
$$c_{\text{air}} = \sqrt{k g_c R T}$$
$$= \sqrt{(1.4)(32.2)(53.3)(460 + 70)}$$
$$= \boxed{1128.5 \text{ ft/sec}}$$

$$E_{\text{steel}} = 29 \times 10^6 \text{ psi} \quad [\text{soft}]$$
$$\rho_{\text{steel}} = 0.283 \text{ lbm/in}^3$$

$$c_{\text{steel}} = \sqrt{\left[\frac{\left(29 \times 10^6 \ \dfrac{\text{lbf}}{\text{in}^2}\right)\left(32.2 \ \dfrac{\text{ft}}{\text{sec}^2}\right)}{\left(0.283 \ \dfrac{\text{lbm}}{\text{in}^3}\right)\left(12 \ \dfrac{\text{in}}{\text{ft}}\right)}\right]\left(\frac{g_c}{g}\right)}$$

$$= \boxed{16{,}580 \text{ ft/sec}}$$

$$E = 320 \times 10^3 \text{ lbf/in}^2$$
$$\rho = 62.3 \text{ lbm/ft}^3$$

$$c_{\text{water}} = \sqrt{\frac{(320 \times 10^3)(144)(32.2)}{62.3}}$$
$$= \boxed{4880 \text{ ft/sec}}$$

6. $R_{\text{He}} = 386.3$

$$\rho = \frac{p}{RT} = \frac{\left(14.7 \ \dfrac{\text{lbf}}{\text{in}^2}\right)\left(144 \ \dfrac{\text{in}^2}{\text{ft}^2}\right)}{\left(386.3 \ \dfrac{\text{ft-lbf}}{\text{lbm-°R}}\right)(460 + 600)°\text{R}}$$

$$= \boxed{0.00517 \text{ lbm/ft}^3}$$

7.
$$h_1 = 1187.2 \text{ BTU/lbm}$$

From the Mollier diagram,

$$h_2 = 953 \text{ BTU/lbm} \quad [\text{at 3 psia, } s_2 = s_1]$$
$$h_2' = 1022 \text{ BTU/lbm}$$
$$\eta_s = \frac{1187.2 - 1022}{1187.2 - 953}$$
$$= \boxed{0.705}$$

8. $(470 \text{ BTU}) \left(2.93 \times 10^{-4} \, \dfrac{\text{kW-hr}}{\text{BTU}}\right)$

$$= \boxed{0.1377 \text{ kwh}}$$

9. The air mass is

$$m = \frac{p_1 V_1}{R T_1}$$

$$= \frac{\left(14.7 \, \dfrac{\text{lbf}}{\text{in}^2}\right)\left(144 \, \dfrac{\text{in}^2}{\text{ft}^2}\right)(8 \text{ ft}^3)}{\left(53.3 \, \dfrac{\text{ft-lbf}}{\text{lbm-}^\circ\text{R}}\right)(460 + 180)^\circ\text{R}}$$

$$= 0.4964 \text{ lbm}$$

For a constant pressure process,

$$W = mR(T_2 - T_1)$$

$$= (0.4964 \text{ lbm})\left(53.3 \, \frac{\text{ft-lbf}}{\text{lbm-}^\circ\text{R}}\right)$$

$$\times (100^\circ\text{F} - 180^\circ\text{F})$$

$$= \boxed{-2116.6 \text{ ft-lbf}}$$

(This is negative because work is done on the system.)

10. Assume that the building needs 3×10^5 ft³/min of 75°F air.

$$\dot{m}_{\text{air}} = \frac{pV}{RT} = \frac{(14.7)(144)(3 \times 10^5)}{(53.3)(460 + 75)}$$

$$= 2.227 \times 10^4 \text{ lbm/hr}$$

This is a constant pressure process, so

$$Q = \dot{m} c_p \Delta T$$

$$= \frac{\left(2.227 \times 10^4 \, \dfrac{\text{lbm}}{\text{hr}}\right)\left(0.241 \, \dfrac{\text{BTU}}{\text{lbm-}^\circ\text{R}}\right)}{3600 \, \dfrac{\text{sec}}{\text{hr}}}$$

$$\qquad\qquad \times (75^\circ\text{F} - 35^\circ\text{F})$$

$$= 59.63 \text{ BTU/sec}$$

For the water,

$$Q = \dot{m}_w c_p \Delta T = V \rho c_p \Delta T$$

The density of water at 165°F is ≈ 61 lbm/ft³.

$$59.63 = \dot{V}_{\text{gpm}} \left(0.002228 \, \frac{\text{ft}^3}{\text{sec-gpm}}\right)\left(61 \, \frac{\text{lbm}}{\text{ft}^3}\right)$$

$$\times \left(1 \, \frac{\text{BTU}}{\text{lbm-}^\circ\text{F}}\right)(180^\circ\text{F} - 150^\circ\text{F})$$

$$\dot{V}_{\text{gpm}} = \boxed{14.63 \text{ gpm}}$$

Concentrates

1. $(5000 \text{ kW})\left(3412.9 \, \dfrac{\text{BTU}}{\text{kW-hr}}\right)$

$$= 1.706 \times 10^7 \text{ BTU/hr}$$

T_{sat} for 200 psia steam is 381.8°F, so this steam is at

$$381.8 + 100 = 481.8^\circ\text{F}$$

$$h_1 \approx 1258 \text{ BTU/lbm}$$

From the Mollier diagram, assuming isentropic expansion,

$$h_2 \approx 868 \text{ BTU/lbm}$$

$$\Delta h = 1258 - 868 = 390 \text{ BTU/lbm}$$

The steam flow rate is

$$\dot{m} = \frac{1.706 \times 10^7 \, \dfrac{\text{BTU}}{\text{hr}}}{390 \, \dfrac{\text{BTU}}{\text{lbm}}}$$

$$= 4.374 \times 10^4 \text{ lbm/hr}$$

The water rate is

$$\text{WR} = \frac{\dot{m}}{\text{kW}} = \frac{4.374 \times 10^4 \, \dfrac{\text{lbm}}{\text{hr}}}{5000 \text{ kW}}$$

$$= \boxed{8.749 \text{ lbm/kW-hr}}$$

When the load decreases, if the steam flow is reduced accordingly, there will be no loss of energy. However, if the steam is throttled to reduce the availability,

$$\text{loss} = \left(\tfrac{1}{2}\right)(h_1 - h_2) = \left(\tfrac{1}{2}\right)(390)$$

$$= \boxed{195 \text{ BTU/lbm}}$$

2. From the Mollier diagram, assuming isentropic expansion,

$$h_1 \approx 1390 \text{ BTU/lbm}$$

$$h_2 \approx 935 \text{ BTU/lbm}$$

The adiabatic heat drop is

$$h_1 - h_2 = 1390 - 935$$

$$= \boxed{455 \text{ BTU/lbm}}$$

3.

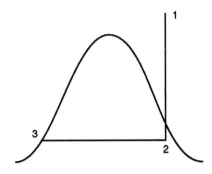

The water rate is

$$\text{WR} = \frac{\dot{m}}{P}$$

$$P = \frac{\dot{m}}{\text{WR}}$$

$$\dot{m} = \frac{P\left(3412.9\ \frac{\text{BTU}}{\text{kW-hr}}\right)}{\Delta h}$$

$$\Delta h = \frac{3412.9}{\text{WR}} = \frac{3412.9}{20}$$

$$= 170.65\ \text{BTU/lbm}$$

The actual steam flow is

$$\dot{m} = P\,(\text{WR}) = (750)(20)$$

$$= 15{,}000\ \text{lbm/hr}$$

$$p_1 = 150\ \text{psig} = 164.7\ \text{psia} \quad [\text{say } 165]$$

For 165 psia steam,

$$T_\text{sat} = 366°\text{F}$$

$$T_1 = T_\text{sat} + T_\text{superheat} = 366°\text{F} + 50°\text{F}$$

$$= 416°\text{F}$$

From the Mollier diagram,

$$h_1 \approx 1226$$

$$p_2 = (26\ \text{in Hg})\left(0.491\ \frac{\text{lbf}}{\text{in}^3}\right)$$

$$= 12.77\ \text{psia}$$

$$h_2 = 1226 - 170.65$$

$$= 1055.4\ \text{BTU/lbm}$$

$$h_{f,3} \approx 171.6\ \text{BTU/lbm} \quad [\text{interpolated}]$$

The heat removal is

$$\left(1055.4 - 171.6\ \frac{\text{BTU}}{\text{lbm}}\right)\left(15{,}000\ \frac{\text{lbm}}{\text{hr}}\right)$$

$$= 1.326 \times 10^7\ \text{BTU/hr}$$

The saturation temperature corresponding to $p_2 = 12.77$ psi is

$$T_2 = T_3 \approx 204°\text{F}$$

Assuming the water and steam leave in thermal equilibrium, the cooling water balance is

$$c_p = 0.999\ \text{BTU/lbm-°F}$$

$$q_\text{in} = \dot{m}c_p\Delta T$$

$$1.326 \times 10^7 = \dot{m}_w(0.999)(204°\text{F} - 65°\text{F})$$

$$\dot{m}_w = 9.55 \times 10^4\ \text{lbm/hr}$$

4. (a) T_sat for 4.45 psia steam $\approx 157°\text{F}$. Assume counter-flow operation to calculate ΔT_m.

81°F　　water　　150°F, 1100 BTU/lbm
A　　　boiling　　B
　　　steam
　　condensing
157°F, saturated liquid　　157°F, saturated vapor

The average water temperature is

$$\left(\tfrac{1}{2}\right)(81 + 150) = 115.5$$

$$c_p \approx 0.999\ \text{BTU/lbm-°F}$$

Assuming 81°F saturated water,

$$h_{w,1} \approx 49\ \text{BTU/lbm}$$

The heat transferred to the water is

$$Q = \dot{m}\Delta h$$

$$= \left(332{,}000\ \frac{\text{lbm}}{\text{hr}}\right)\left(1100\ \frac{\text{BTU}}{\text{lbm}} - 49\ \frac{\text{BTU}}{\text{lbm}}\right)$$

$$= 3.489 \times 10^8\ \text{BTU/hr}$$

$$\Delta T_m = \frac{(157 - 81) - (157 - 150)}{\ln\left(\dfrac{157 - 81}{157 - 150}\right)}$$

$$= 28.93$$

Since $T_\text{steam,A} = T_\text{steam,B}$, the correction factor for ΔT_m (F_c) is 1.

$$U = \frac{Q}{A\Delta T_m} = \frac{3.489 \times 10^8\ \dfrac{\text{BTU}}{\text{hr}}}{(1850\ \text{ft}^2)(28.93°\text{F})}$$

$$\boxed{= 6519\ \text{BTU/hr-ft}^2\text{-°F}}$$

(b) The enthalpy of saturated 4.45 psia steam is

$$h_1 \approx 1128\ \text{BTU/lbm}$$

The enthalpy of saturated 4.45 psia water is

$$h_2 \approx 125 \text{ BTU/lbm}$$

The enthalpy change is

$$h_1 - h_2 = 1128 - 125 = 1003 \text{ BTU/lbm}$$

From an energy balance with the water,

$$3.489 \times 10^8 = 1003\dot{m} \qquad Q = \dot{m}\,(\Delta h)$$

$$\dot{m} = 3.479 \times 10^5 \text{ lbm/hr steam}$$

The extraction rate is

$$\dot{m}h_1 = \left(3.479 \times 10^5 \frac{\text{lbm}}{\text{hr}}\right)\left(1128 \frac{\text{BTU}}{\text{lbm}}\right)$$

$$= \boxed{3.92 \times 10^8 \text{ BTU/hr}}$$

5.

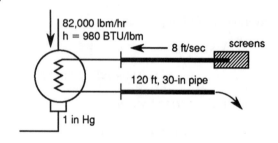

82,000 lbm/hr
h = 980 BTU/lbm

8 ft/sec   screens

120 ft, 30-in pipe

1 in Hg

(a) For X pipe,

$$D_i = 2.4167 \text{ ft}$$

$$A_i = 4.5869 \text{ ft}^2$$

$\epsilon = 0.0002$ ft for steel.

$$\frac{\epsilon}{D} = \frac{0.0002}{2.4167} = 0.000083$$

For fully turbulent flow,

$$f \approx 0.012 \qquad h_f = \frac{(f)\,v^2}{2Dg}$$

friction head
$$h_f = \frac{(0.012)(120)(8)^2}{(2)(2.4167)(32.2)} = 0.59 \text{ ft}$$

(This will vary depending on the value of f chosen.)

The screen loss is 6 in or 0.5 ft. The total head added by a coolant pump (not shown), including velocity head but not including losses inside the condenser, is

$$0.59 + 0.5 + \frac{(8)^2}{(2)(32.2)} = \boxed{2.08 \text{ ft}}$$

Velocity head
$$h_v = \frac{v^2}{2g}$$

(b) The water flow can be found from the pipe flow velocity.

$$vA = \frac{\left(8\,\frac{\text{ft}}{\text{sec}}\right)(4.5869 \text{ ft}^2)}{0.002228\,\frac{\text{ft}^3}{\text{sec-gpm}}}$$

$$= \boxed{16,470 \text{ gpm}}$$

Since this flow rate does not use the 10°F information, check with a heat balance. $849 \frac{lbm}{ft^3}$

$$(1 \text{ in Hg})\left(\underset{\rho_{mercury} =}{0.491}\frac{\text{lbm}}{\text{in}^3}\right) \approx 0.5 \text{ psi}$$

$$h_{\text{sat liquid at 0.5 psi}} = 47.6 \text{ BTU/lbm}$$

For the steam,

$$Q_{\text{out}} = \dot{m}\Delta h = \frac{\left(82,000\,\frac{\text{lbm}}{\text{hr}}\right)\left(980 - 47.6\,\frac{\text{BTU}}{\text{lbm}}\right)}{60\,\frac{\text{min}}{\text{hr}}}$$

$$= 1.274 \times 10^6 \text{ BTU/min}$$

For the water,

$$1.274 \times 10^6\,\frac{\text{BTU}}{\text{min}} = \dot{m}c_p\Delta T$$

$$= \dot{m}\left(1\,\frac{\text{BTU}}{\text{lbm-°F}}\right)(10°\text{F})$$

$$\dot{m} = 1.274 \times 10^5 \text{ lbm/min}$$

$$\frac{\dot{m}}{\rho} = \frac{\left(1.274 \times 10^5\,\frac{\text{lbm}}{\text{min}}\right)\left(7.48\,\frac{\text{gal}}{\text{ft}^3}\right)}{62.4\,\frac{\text{lbm}}{\text{ft}^3}}$$

$$= \boxed{1.527 \times 10^4 \text{ gpm}}$$

Both answers are reasonably consistent.

6. $$h_1 = 28.06 \text{ BTU/lbm}$$

$$h_2 = 1150.4 \text{ BTU/lbm}$$

$$\Delta h = 1150.4 - 28.06 = 1122.3 \text{ BTU/lbm}$$

$$Q = \dot{m}\Delta h = (100)(1122.3)$$

$$= 1.122 \times 10^5 \text{ BTU/hr}$$

$$\frac{\left(1.122 \times 10^5\,\frac{\text{BTU}}{\text{hr}}\right)\left(0.2930\,\frac{\text{W-hr}}{\text{BTU}}\right)}{1000\,\frac{\text{W}}{\text{kW}}} = 32.87 \text{ kW}$$

PROFESSIONAL PUBLICATIONS, INC. ● Belmont, CA

$$\text{cost} = \frac{(32.87\,\text{kW})\left(\dfrac{\$0.04}{\text{kW-hr}}\right)}{1-0.35}$$

$$= \boxed{\$2.02/\text{hr}}$$

7. $p = 25 + 14.7 = 39.7$ psia [say 40 psia]

At 60°F,

$$h_1 = 28.06\ \text{BTU/lbm}$$

$$h_2 = 236.03 + (0.98)(933.7)$$

$$= 1151.06\ \text{BTU/lbm}$$

$$\Delta h = 1151.06 - 28.06 = 1123\ \text{BTU/lbm}$$

$$Q = \dot{m}\Delta h = \frac{\left(250\ \dfrac{\text{lbm}}{\text{hr}}\right)\left(1123\ \dfrac{\text{BTU}}{\text{lbm}}\right)}{60\ \dfrac{\text{min}}{\text{hr}}}$$

$$= 4679.2\ \text{BTU/min}$$

Find the volume of gas used at standard conditions for a heating gas (60°F).

$$\dot{V} = \left(13.5\ \frac{\text{ft}^3}{\text{min}}\right)\left(\frac{460+60}{460+80}\right)$$

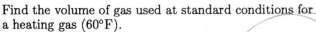

(handwritten annotation) $\dfrac{P_1 V_1}{T_1} = \dfrac{P_2 V_2}{T_2}$ converting to STP

$$\times \left[\frac{(30.2)(0.491)+(4)(0.0361)}{14.7}\right]$$

$$= 13.24\ \text{SCFM}$$

$$\eta = \frac{4679.2\ \dfrac{\text{BTU}}{\text{min}}}{\left(13.24\ \dfrac{\text{ft}^3}{\text{min}}\right)\left(550\ \dfrac{\text{BTU}}{\text{ft}^3}\right)}$$

$$= 0.643$$

8. The tangential blade speed is

$$v_b = \frac{\pi\left(\dfrac{18\ \text{in}}{12\ \dfrac{\text{in}}{\text{ft}}}\right)(12{,}000\ \text{rpm})}{60\ \dfrac{\text{sec}}{\text{min}}}$$

$$= 942.5\ \text{ft/sec}$$

The jet speed is

$$v = \frac{942.5}{0.4} = 2356.3\ \text{ft/sec}$$

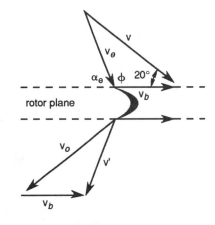

rotor plane

The relative jet speed is

$$v_e = \sqrt{(2356.3)^2 + (942.5)^2 - (2)(2356.3)(942.5)(\cos 20°)}$$

$$= 1505.6\ \text{ft/sec}$$

The angle ϕ can be found using the law of sines.

$$\frac{1505.5}{\sin 20°} = \frac{2356.2}{\sin \phi}$$

$$\phi = 147.6°$$

$$\alpha_e = 180° - \phi = 32.4°$$

The actual steam exit velocity is

$$v' = \sqrt{(942.5)^2 + (1505.6)^2 - (2)(942.5)(1505.6)(\cos 32.4°)}$$

$$= 871.1\ \text{ft/sec}$$

The reduction in kinetic energy is

$$\Delta E_k = \frac{(2356.3)^2 - (871.1)^2}{(2)(32.2)\left(778\ \dfrac{\text{ft-lbf}}{\text{BTU}}\right)} = 95.67\ \text{BTU/lbm}$$

9. *step 1:* Determine the actual gravimetric analysis of the coal as fired. Use the successive deletion method on a per-pound basis. 1 lbm of coal contains 0.02 lbm moisture, leaving 0.98 lbm coal. Of this, 5% is ash, so the weight of ash is $(0.05)(0.98) = 0.049$ lbm. The remainder $(0.98 - 0.049 = 0.931)$ is assumed to be carbon.

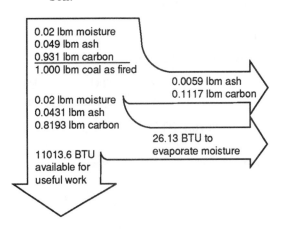

0.02 lbm moisture
0.049 lbm ash
0.931 lbm carbon
1.000 lbm coal as fired

0.0059 lbm ash
0.1117 lbm carbon

0.02 lbm moisture
0.0431 lbm ash
0.8193 lbm carbon

26.13 BTU to evaporate moisture

11013.6 BTU available for useful work

step 2: Determine the ash pit material losses. 12% of the dry coal goes to the ash pit.

$$(0.12)(0.049) = 0.0059 \text{ ash}$$
$$(0.12)(0.931) = 0.1117 \text{ unburned carbon}$$

step 3: Determine what remains.

$$0.02 \text{ lbm moisture}$$
$$0.049 - 0.0059 = 0.0431 \text{ lbm fly ash}$$
$$0.931 - 0.1117 = 0.8193 \text{ lbm carbon}$$

step 4: Determine energy losses. The 0.02 lbm moisture has to be evaporated. Assume the coal is initially at 60°F, and the combustion occurs at 14.7 psia. Assume the combustion products leave at 600°F.

$$h_{60°F} = 28.06 \text{ BTU/lbm}$$
$$h_{600°F, 1 \text{ atm}} = 1334.8$$
$$Q = (0.02)(1334.8 - 28.06) = 26.13 \text{ BTU}$$

step 5: Calculate the heating value of the remaining coal. The heating value is given per pound of coal, not per pound of carbon.

$$\text{weight of coal} = 0.0431 + 0.8193$$
$$= 0.8624 \text{ lbm}$$
$$(0.8624 \text{ lbm}) \left(12{,}800 \ \frac{\text{BTU}}{\text{lbm}}\right) = 11{,}039 \text{ BTU}$$

step 6: Subtract the losses.

$$11{,}039 - 26.13 = 11{,}013 \text{ BTU}$$

step 7: Find the energy required to vaporize steam.

$$h_1 = 87.92 \text{ BTU/lbm}$$
$$100 \text{ psig} \approx 115 \text{ psia}$$
$$h_2 = 1190 \text{ BTU/lbm}$$
$$Q = (8.23 \text{ lbm}) \left(1190 \ \frac{\text{BTU}}{\text{lbm}} - 87.92 \ \frac{\text{BTU}}{\text{lbm}}\right)$$
$$= 9070.1 \text{ BTU}$$

step 8: The combustion efficiency is

$$\eta = \frac{9070.1}{11{,}013} = \boxed{0.824}$$

10. *step 1:* The incoming reactants on a per-pound basis are

$$0.07 \text{ lbm ash}$$
$$0.05 \text{ lbm hydrogen}$$
$$0.05 \text{ lbm oxygen}$$
$$0.83 \text{ lbm carbon}$$

This is an ultimate analysis. Assume that all of the oxygen and 1/8 of the hydrogen are in the form of water. The reactants as compounds are

$$0.07 \text{ lbm ash}$$
$$0.05625 \text{ lbm moisture}$$
$$0.04375 \text{ lbm hydrogen}$$
$$0.83 \text{ lbm carbon}$$

The air is 23.15% oxygen by weight, so other reactants are

$$(0.2315)(26) = 6.019 \text{ lbm oxygen}$$
$$(0.7685)(26) = 19.981 \text{ lbm nitrogen}$$

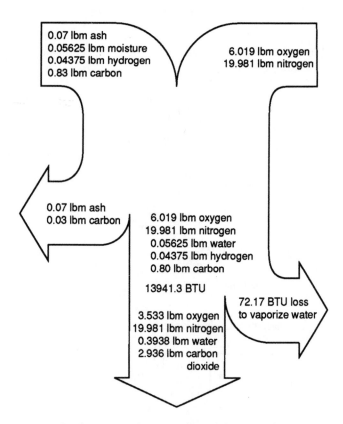

step 2: Assume ash pit material losses. Assume a 0.1 lbm loss, which includes all of the ash.

$$0.07 \text{ lbm ash}$$
$$0.03 \text{ lbm unburned coal}$$

step 3: Determine what remains.

step 4: Determine the energy loss in vaporizing the moisture. Assume the coal is initially at 60°F and that combustion occurs at 14.7 psia.

$$h_{60°F} = 28.06 \text{ BTU/lbm}$$

$$h_{550°F} = 1311 \text{ BTU/lbm}$$

$$Q = (0.05625)(1311 - 28.06)$$

$$= 72.17 \text{ BTU}$$

step 5: Calculate the heating value of the remaining fuel components.

$$Q_{\text{carbon}} = (0.80 \text{ lbm}) \left(14{,}093 \, \frac{\text{BTU}}{\text{lbm}} \right)$$

$$= 11{,}274.4 \text{ BTU}$$

$$Q_{\text{hydrogen}} = (0.04375)(60{,}958)$$

$$= 2666.9 \text{ BTU}$$

$$Q_{\text{total}} = 11{,}274.4 + 2666.9 = 13{,}941.3 \text{ BTU}$$

The heating value after the coal moisture is evaporated is

$$13{,}941.3 - 72.17 = 13{,}869 \text{ BTU}$$

step 6: Determine the combustion products. The carbon needs

$$(0.8)(2.67) = 2.136 \text{ lbm oxygen} \qquad \text{C+O}_2 = \text{CO}_2$$

The carbon produces

$$(0.8)(3.67) = 2.936 \text{ lbm carbon dioxide}$$

The hydrogen needs

$$(0.04375)(8) = 0.35 \text{ lbm oxygen} \qquad \text{H}_2 + \text{O}_2 = \text{H}_2\text{O}$$

The hydrogen produces

$$(0.04375)(9) = 0.3938 \text{ lbm water}$$

The remaining oxygen is

$$6.019 - 2.136 - 0.35 = 3.533 \text{ lbm}$$

MOLECULAR WEIGHT RATIOS

step 7: The gaseous products must be heated from 70°F to 550°F. The average temperature is

$$\left(\tfrac{1}{2}\right)(70°F + 550°F) = 310°F \qquad [770°R]$$

c_p for these gases is

	c_p
oxygen	0.228
nitrogen	0.252
water	0.460
carbon dioxide	0.225

$$Q_{\text{heating}} = [(3.533)(0.228) + (19.981)(0.252)$$
$$+ (0.3938)(0.460) + (2.936)(0.225)]$$
$$\times (550°F - 70°F)$$

$$= 3207.6 \text{ BTU}$$

step 8: The percent loss is

$$\frac{3207.6 + 72.17}{13{,}941.3} = 0.235 \quad (23.5\%)$$

Timed

1.

15 lbm/hr
140° F

0.25 hp

15 psia

The drill power is

$$(0.25 \text{ hp}) \left(2544.9 \, \frac{\text{BTU}}{\text{hp-hr}} \right) = 636.23 \text{ BTU/hr}$$

$$T = 460 + 140°F = 600°R$$

$$h_1 = 143.47 \text{ BTU/lbm}$$

$$p_{r,1} = 2.005$$

$$\phi_1 = 0.62607$$

$$W_{\text{actual}} = \dot{m}(h_1 - h_{2,\text{actual}})$$

$$= \dot{m}\eta(h_1 - h_{2,\text{ideal}})$$

$$h_{2,\text{ideal}} = h_1 - \frac{W}{\dot{m}\eta}$$

$$= 143.47 \, \frac{\text{BTU}}{\text{lbm}} - \frac{636.23 \, \dfrac{\text{BTU}}{\text{hr}}}{\left(15 \, \dfrac{\text{lbm}}{\text{hr}} \right)(0.60)}$$

$$= 72.778 \text{ BTU/lbm}$$

From Keenan and Kayes *Gas Tables*, for $h = 72.778$ BTU/lbm,

$$T = 305°R$$

$$p_{r,2} = 0.18851$$

(a) Due to the irreversibility of the expansion,

$$h_2' = 143.47 - (0.60)(143.47 - 72.778)$$

$$= 101.05 \text{ BTU/lbm}$$

Searching the *Gas Tables* again gives

$$T_2' = 423°R$$

$$\phi_2 = 0.54228$$

(b) Since $p_1/p_2 = p_{r,1}/p_{r,2}$,

$$p_1 = (15 \text{ psia})\left(\frac{2.005}{0.18851}\right) = 159.5 \text{ psia}$$

The ratios of pressure are valid only for isenthropic processes; therefore, $p_{r,2}$ must be used, not $p_{r,2}'$.

$$s_2 - s_1 = 0.54228 - 0.62607 - \left(\frac{53.3}{778}\right)\ln\left(\frac{15}{159.5}\right)$$

$$= 0.07816 \text{ BTU/lbm-°R}$$

2.

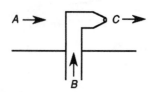

For the steam,

$$T_A = 600°F$$
$$p_A = 200 \text{ psig} \approx 215 \text{ psia}$$
$$h_A = 1321 \text{ BTU/lbm}$$
$$s_A = 1.6680 \text{ BTU/lbm-°R}$$

For the water,

$$T_B = 82°F$$
$$h = 50.03 \text{ BTU/lbm}$$

The correction for compression is ≈ 0.55 BTU/lbm.

$$h_B = 50.03 + 0.55 = 50.58 \text{ BTU/lbm}$$
$$s_B = 0.0969 \text{ BTU/lbm-°R}$$

From an energy balance,

$$h_A m_A + h_B m_B = h_C(m_A + m_B)$$
$$h_C = \frac{(1000)(1321) + (50)(50.58)}{1000 + 50}$$
$$= 1260.5 \text{ BTU/lbm}$$

Some of this energy goes into kinetic energy.

$$E_k = \frac{\left(2000 \frac{\text{ft}}{\text{sec}}\right)^2}{(2)\left(32.2\frac{\text{ft}}{\text{sec}^2}\right)\left(778\frac{\text{ft-lbf}}{\text{BTU}}\right)}$$
$$= 79.8 \text{ BTU}$$

$$h_C' = 1260.5 - 79.8 = 1180.7 \text{ BTU/lbm}$$

For 100 psia steam,

$$h_f = 298.4 \text{ BTU/lbm}$$
$$h_{fg} = 888.8 \text{ BTU/lbm}$$
$$x = \frac{1180.7 - 298.4}{888.8} = 0.993$$

T_C is the saturation temperature for 100 psia steam, 327.8°F.

$$s_C = 0.4740 + (0.993)(1.1286)$$
$$= 1.5947 \text{ BTU/lbm-°R}$$

The entropy production is

$$\left(1050\frac{\text{lbm}}{\text{hr}}\right)\left(1.5947\frac{\text{BTU}}{\text{lbm-°R}}\right) - (1000)(1.6680)$$
$$- (50)(0.0969)$$
$$= 1.59 \text{ BTU/hr-°R}$$

3. (a) $R = 11.9$. Assume the tank is originally at 70°F.
$$m = \frac{pV}{RT} = \frac{(20)(144)(100)}{(11.9)(460°F + 70°F)}$$
$$= \boxed{45.66 \text{ lbm}}$$

(b) The critical properties for xenon are
$$p_c = 855.3 \text{ psia}$$
$$T_c = 521.9°R$$

At the tank conditions,
$$p_r = \frac{3800}{855.3} = 4.44$$
$$T_r = \frac{530}{521.9} = 1.02$$
$$z = 0.6$$
$$m = \frac{pV}{zRT} = \frac{(3800)(144)(100)}{(0.6)(11.9)(530)}$$
$$= 14{,}460 \text{ lbm}$$
$$\dot{m} = 14{,}460 - 45.66$$
$$= \boxed{14{,}414 \text{ lbm/hr}}$$

(c) Since a heat exchanger was mentioned, assume isothermal compression.

$$W = p_1 V_1 \ln \left(\frac{V_2}{V_1} \right)$$

$$= mRT_1 \ln \left(\frac{p_1}{p_2} \right)$$

$$= \frac{(14{,}414)(11.9)(460+70)\ln\left(\dfrac{20}{3800}\right)}{(778)(3413)}$$

$$= -180 \text{ kW-hr}$$

$$\text{cost} = (0.045)(180) = \boxed{\$8.10}$$

This calculation assumes the compressor efficiency is constant over the entire range of receiver pressures.

Notice that Z does not affect the work equation. Z affects only the mass flow rate.

4. Choose the control volume to include the air outside the tank that is pushed into the tank, as well as the tank volume.

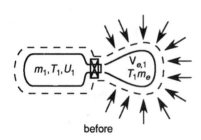

before

after

$m_1 =$ air in tank when evacuated

$$= \frac{p_1 V_1}{RT_1}$$

$$= \frac{(1 \text{ psia})\left(144 \dfrac{\text{in}^2}{\text{ft}^2}\right)(20 \text{ ft}^3)}{\left(53.3 \dfrac{\text{ft-lbf}}{\text{lbm-}^\circ\text{R}}\right)(70+460)^\circ\text{R}}$$

$$= 0.102 \text{ lbm}$$

Assume $T_2 = 80^\circ\text{F}$.

$$m_1 + m_e = \frac{(14.7)(144)(20)}{(53.3)(80+460)}$$

$$= 1.471 \text{ lbm}$$

$$m_e \approx 1.471 - 0.102$$

$$= 1.369 \text{ lbm}$$

The initial volume of the external air is

$$V_{e,1} \approx \frac{mRT}{p} = \frac{(1.369)(53.3)(70+460)}{(14.7)(144)}$$

$$= 18.27 \text{ ft}^3$$

This is a closed system. The first law of thermodynamics for closed systems is

$$Q = \Delta U + W = U_2 - U_1 + W$$

Since the system is adiabatic, $Q = 0$.

$$W_{\text{ext}} = \Delta U$$

The external work is a constant pressure process. W_{ext} is negative because the surroundings do work on the system.

$$W_{\text{ext}} = p(V_{e,2} - V_{e,1})$$

$$= \frac{(14.7)(144)(0 - 18.27)}{778 \dfrac{\text{ft-lbf}}{\text{BTU}}}$$

$$= -49.71 \text{ BTU}$$

Since $\Delta U = mc_v \Delta T$ for the air and $c_p = c_v$ for the tank material,

$$W_{\text{ext}} = [(m_1 + m_e)c_v + m_{\text{tank}}c_p](T_1 - T_2)$$

$$-49.71 = [(1.471)(0.171) + (40)(0.11)](70 - T_2)$$

$$T_2 = \boxed{80.69^\circ\text{F}}$$

This is close enough to the assumed value of T_2 that a second iteration is unnecessary.

5. Assume that pressures are low enough to ignore compressibility. Calculate the flow rates.

$$\dot{m}_C = \frac{pV}{RT} = \frac{(80)(144)(100)}{(53.3)(460+85)}$$

$$= 39.66 \text{ lbm/min}$$

$$\dot{m}_D = \frac{(85)(144)(120)}{(53.3)(460+80)}$$

$$= 51.03 \text{ lbm/min}$$

$$\dot{m}_E = 8 \text{ lbm/min} \quad \text{[given]}$$

$$\dot{m}_{\text{total}} = 39.66 + 51.03 + 8 = 98.69 \text{ lbm/min}$$

The input from compressor A is

$$\dot{m}_A = \frac{(14.7)(144)(600)}{(53.3)(460+80)}$$

$$= 44.13 \text{ lbm/min}$$

$$\dot{m}_B = 98.69 - 44.13 = 54.56 \text{ lbm/min}$$

$$\dot{V}_B = \left(\frac{54.56}{44.13}\right)(600)$$

$$= \boxed{742 \text{ ft}^3/\text{min}}$$

6.

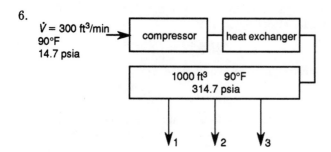

Assume steady flow and constant properties.

$$\dot{m} = \frac{p\dot{V}}{RT}$$

$$= \frac{(14.7)(144)(300)}{(53.3)(550)}$$

$$= 21.7 \text{ lbm/min}$$

Mass initially in storage:

$$m = \frac{pV}{RT} = \frac{(314.7)(144)(1000)}{(53.3)(550)}$$

$$= 1545.9 \text{ lbm}$$

Assuming that each tool operates at its minimum pressure, the mass leaving the system is

Tool 1:

$$\dot{m}_{\text{out},1} = \frac{(104.7)(144)(40)}{(53.3)(550)}$$

$$= 20.57 \text{ lbm/min}$$

Tool 2:

$$\dot{m}_{\text{out},2} = \frac{(64.7)(144)(15)}{(53.3)(545)}$$

$$= 4.81 \text{ lbm/min}$$

Tool 3:

$$\dot{m}_{\text{out},3} = 6 \text{ lbm/min}$$

The total mass flow rate leaving the system is

$$\dot{m}_{\text{out}} = 20.57 + 4.81 + 6 = 31.38 \text{ lbm/min}$$

The critical pressure is 90 psig = 104.7 psia. The mass in the tank when critical pressure is achieved is

$$\dot{m}_{\text{tank,critical}} = \frac{(104.7)(144)(1000)}{(53.3)(550)}$$

$$= 514.3 \text{ lbm}$$

The net flow rate of air to the tank is

$$\dot{m}_{\text{in}} - \dot{m}_{\text{out}} = 21.7 - 31.38$$

$$= -9.68 \text{ lbm/min} \quad \begin{bmatrix} \text{rate of depletion} \\ \text{of tank} \end{bmatrix}$$

The amount to be depleted is

$$m_{\text{initial}} - m_{\text{final}} = 1545.9 - 514.3 = 1031.6 \text{ lbm}$$

The time the system can run is

$$\frac{1031.6 \text{ lbm}}{9.68 \dfrac{\text{lbm}}{\text{min}}} = \boxed{106.57 \text{ min} \quad [1.78 \text{ hr}]}$$

7.

$$T_1 = 540°\text{F} \qquad T_2 = 2000°\text{R}$$

$$= 1000°\text{R} \qquad p_2 = 80 \text{ psia}$$

$$p_1 = 100 \text{ psia} \qquad T_o = 560°\text{R}$$

$$W_{\text{max}} = \Phi_1 - \Phi_2$$

$$\Phi = h - T_o s$$

Since pressures are low (< 300 psia) and temperatures are much higher than 235.8°R, use an air table. For $T_1 = 1000°\text{R}$,

$$h_1 = 240.98 \text{ BTU/lbm}$$

$$\phi_1 = 0.75042 \text{ BTU/lbm-°R}$$

$$T_2 = 2000°\text{R}$$

$$h_2 = 504.71 \text{ BTU/lbm}$$

$$\phi_2 = 0.93205 \text{ BTU/lbm-°R}$$

$$W_{\text{max}} = h_1 - h_2 + (s_2 - s_1)T_o$$

For no pressure drop,

$$s_2 - s_1 = \phi_2 - \phi_1$$

$$W_{\text{max}} = 240.98 - 504.71 + (0.93205 - 0.75042)(560)$$

$$= -162.02 \text{ BTU/lbm}$$

With a pressure drop from 100 to 80 psia,

$$s_2 - s_1 = \phi_2 - \phi_1 - \left(\frac{R}{J}\right) \ln\left(\frac{p_2}{p_1}\right)$$

$$W_{\max,p\,\text{loss}} = 240.98 - 504.71$$

$$+ \left[0.93205 - 0.75042 \right.$$

$$\left. - \left(\frac{53.3}{778}\right) \times \ln\left(\frac{80}{100}\right) \right] (560)$$

$$= -153.46 \text{ BTU/lbm}$$

$$\frac{(-153.46) - (-162.02)}{-162.02} = -0.0528$$

$$= \boxed{5.28\% \text{ loss}}$$

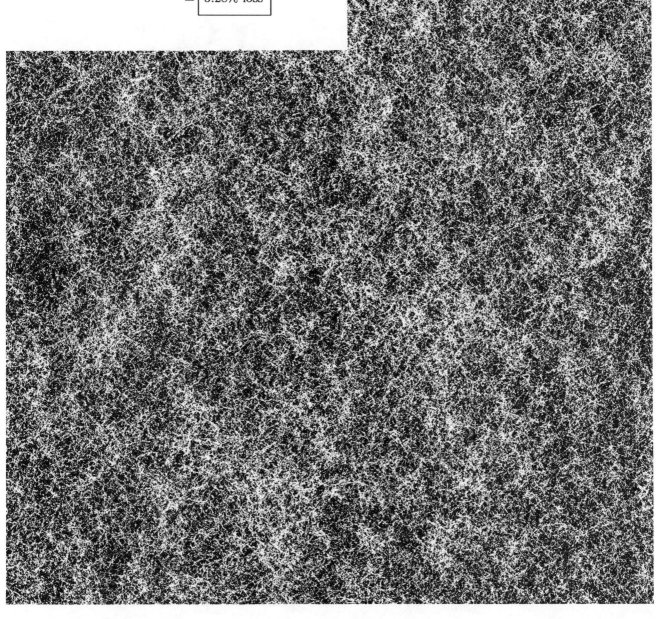

PROFESSIONAL PUBLICATIONS, INC. ● Belmont, CA

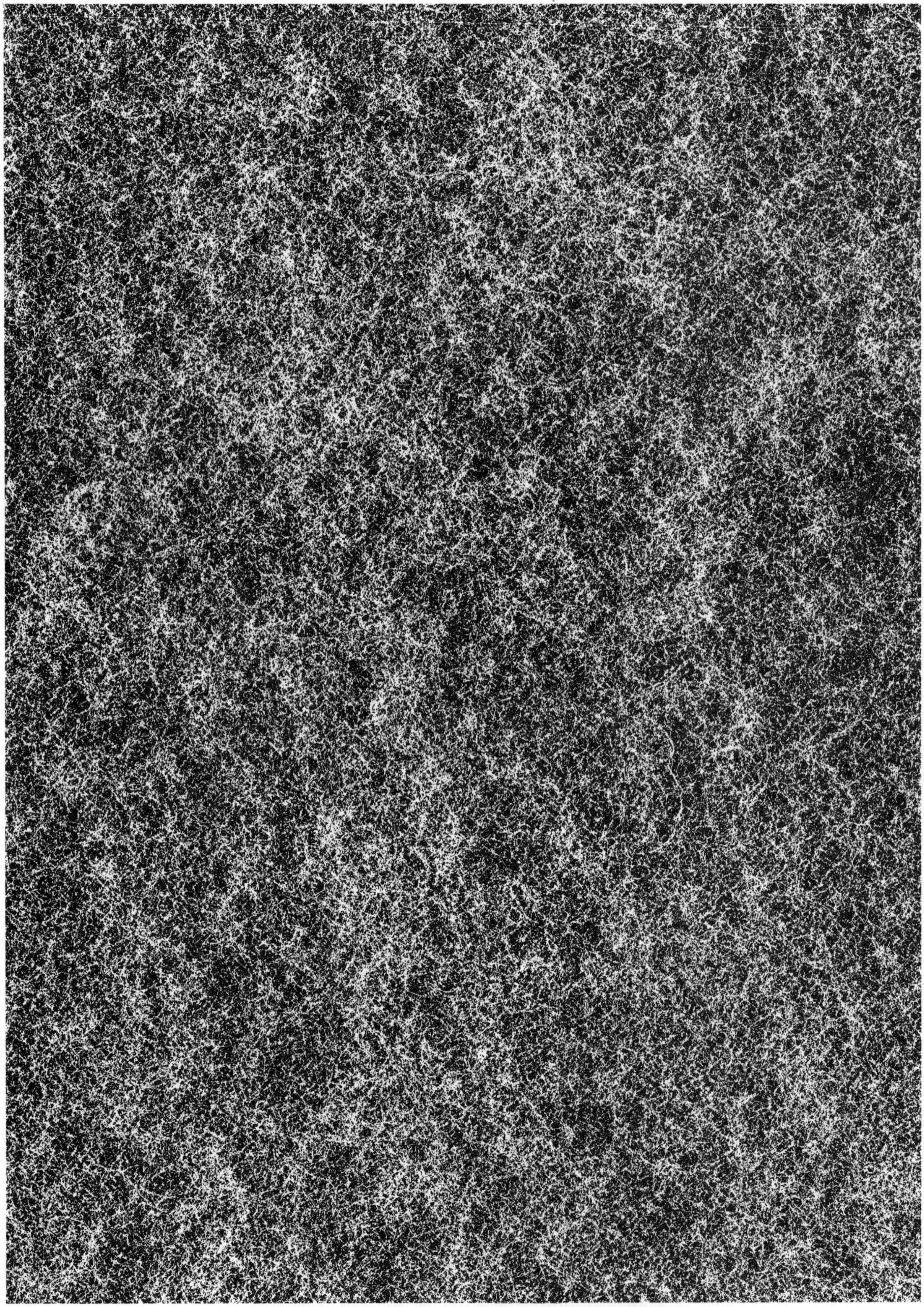

POWER CYCLES

Warmups

1.
$$\eta_{\text{th}} = \frac{(650 + 460) - (100 + 460)}{650 + 460}$$

$$= \boxed{0.495 \ (49.5\%)}$$

2.
$$\text{COP} = \frac{700 + 460}{(700 + 460) - (40 + 460)} = \boxed{1.76}$$

3. Refer to Fig. 7.12.

At a,

$$T_a = 650°F$$
$$h_a = 696.4 \ \text{BTU/lbm} \quad \text{[interpolated]}$$
$$s_a = 0.8833 \ \text{BTU/lbm-°F}$$

At b,

$$T_b = 650°F$$
$$h_b = 1117.5 \quad \text{[interpolated]}$$
$$s_b = 1.2631$$

At c,

$$T_c = 100°F$$
$$s_c = s_b = 1.2631$$
$$x_c = \frac{1.2631 - 0.1295}{1.8531} = 0.612$$
$$h_c = 67.97 + (0.612)(1037.2)$$
$$= 702.7$$

At d,

$$T_d = 100°F$$
$$s_d = s_a = 0.8833$$
$$x_c = \frac{0.8833 - 0.1295}{1.8531} = 0.407$$
$$h_d = 67.97 + (0.407)(1037.2)$$
$$= 490.1$$

Due to the inefficiencies,

$$h'_c = 1117.5 - (0.9)(1117.5 - 702.7) = 744.2$$

$$h'_a = 490.1 + \frac{696.4 - 490.1}{0.8} = 748.0$$

$$n_{\text{th}} = \frac{(1117.5 - 744.2) - (748.0 - 490.1)}{1117.5 - 748.0}$$

$$= \boxed{0.312 \ (31.2\%)}$$

4.
$$\text{COP} = \frac{5 + 460}{(90 + 460) - (5 + 460)} = 5.47$$

$$\text{power} = \frac{(1 \ \text{ton}) \left(200 \ \dfrac{\text{BTU}}{\text{min-ton}} \right) \left(778 \ \dfrac{\text{ft-lbf}}{\text{BTU}} \right)}{\left(5.47 \ \dfrac{\text{BTU out}}{\text{BTU in}} \right) \left(33{,}000 \ \dfrac{\text{ft-lbf}}{\text{min-hp}} \right)}$$

$$= 0.862 \ \text{hp}$$

$$\text{EER} = \frac{\left(200 \ \dfrac{\text{BTU}}{\text{min}} \right) \left(60 \ \dfrac{\text{min}}{\text{hr}} \right)}{(0.862 \ \text{hp}) \left(745.7 \ \dfrac{\text{W}}{\text{hp}} \right)}$$

$$= \boxed{18.7}$$

5. The specific gravity is

$$\text{SG} = \frac{141.5}{131.5 + 40} = 0.825$$

$$\text{HHV} = 22{,}320 - (3780)(\text{SG})^2$$
$$= 22{,}320 - (3780)(0.825)^2$$

$$= \boxed{19{,}747 \ \text{BTU/lbm}}$$

6. The actual horsepower is

$$\text{power} = \frac{(\text{rpm})(\text{torque in ft-lbf})}{5252}$$

$$= \frac{(200)(600)}{5252} = 22.85 \ \text{hp}$$

The number of power strokes per minute is

$$N = \frac{(2)(200)(2)}{4} = 200$$

The stroke is

$$\frac{18}{12} = 1.5 \ \text{ft}$$

The bore area is

$$\left(\frac{\pi}{4} \right) (10)^2 = 78.54 \ \text{in}^2$$

The ideal horsepower is

$$\text{power} = \frac{(95)(1.5)(78.54)(200)}{33{,}000} = 67.83 \ \text{hp}$$

The friction horsepower is

$$67.83 - 22.85 = \boxed{44.98 \text{ hp}}$$

7.
$$R = \frac{65}{14.7} = 4.42$$

$$\eta_v = 1 - [(4.42)^{\frac{1}{1.33}} - 1](0.07)$$

$$= 0.856$$

The mass of air compressed per minute is

$$\frac{48}{0.856} = \boxed{56.07 \text{ lbm/min}}$$

8.

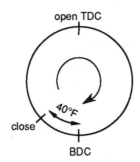

open TDC

40°F

close

BDC

The number of degrees that the valve is open is

$$180° + 40° = 220°$$

The time that the valve is open is

$$\left(\frac{220}{360}\right)\left(\frac{\text{time}}{\text{revolution}}\right) = \left(\frac{220}{360}\right)\left(\frac{60 \frac{\text{sec}}{\text{min}}}{4000 \text{ rpm}}\right)$$

$$= 9.167 \times 10^{-3} \text{ sec}$$

The displacement is

$$\left(\frac{\pi}{4}\right)\left(\frac{3.1}{12}\right)^2\left(\frac{3.8}{12}\right) = 0.0166 \text{ ft}^3$$

The actual incoming volume per intake stroke is

$$V = (0.65)(0.0166) = 0.01079 \text{ ft}^3$$

The area is

$$A = \frac{V}{\text{v}t} = \frac{0.01079 \text{ ft}^3}{\left(100 \frac{\text{ft}}{\text{sec}}\right)(9.167 \times 10^{-3} \text{ sec})}$$

$$= \boxed{0.0118 \text{ ft}^2 \ (1.69 \text{ in}^2)}$$

9. *Method 1:* Use ideal gas relationships.

At 1500°F (1960°R), the ratio of specific hets for air is approximately 1.33.

$$V_2 = \left(10 \frac{\text{ft}^3}{\text{sec}}\right)\left(\frac{200 \text{ psia}}{50 \text{ psia}}\right)^{\frac{1}{1.33}}$$

$$= 28.358 \text{ ft}^3/\text{sec}$$

$$T_2 = (1960°R)\left(\frac{50}{200}\right)^{\frac{1.33-1}{1.33}}$$

$$= \boxed{1379°R}$$

The average temperature is

$$T_{\text{ave}} = \frac{1960°R + 1390°R}{2} = 1675°R$$

At 1675°R, the ratio of specific heats is approximately 1.34. Using this value,

$$V_2 = \boxed{28.138 \text{ ft}^3/\text{sec}}$$

$$T_2 = \boxed{1379°R}$$

At 1675°R, the specific heat can be found from a variety of ways. One way is to look at values of enthalpy in an air table and calculate $c_p = \Delta h/\Delta T$. However, if an air table is available, use Method 2.

$$c_p = a + bT + cT^2 + \frac{d}{\sqrt{T}}$$

$$= 0.2459 + (3.22 \times 10^{-5})(1675°R)$$

$$\quad - (3.74 \times 10^{-9})(1675°R)^2 - \frac{0.833}{\sqrt{1675°R}}$$

$$= 0.269 \text{ BTU/lbm-°R}$$

$$\Delta h = c_p \Delta T$$

$$= \left(0.269 \frac{\text{BTU}}{\text{lbm-°R}}\right)(1960°R - 1379°R)$$

$$= \boxed{156.3 \text{ BTU/lbm} \quad [\text{decrease}]}$$

Method 2: Using air tables at 1960°R,

$$h_1 = 493.64 \text{ BTU/lbm}$$

$$p_{r,1} = 160.37$$

$$\text{v}_{r,1} = 4.527$$

After expansion,

$$p_{r,2} = (160.37)\left(\frac{50}{200}\right) = 40.09$$

Searching the table for this value of $p_{r,2}$,

$$T_2 = \boxed{1375°\text{R}}$$

$$h_2 = 336.39 \text{ BTU/lbm}$$
$$v_{r,1} = 12.721$$

$$V_2 = (10)\left(\frac{12.721}{4.527}\right) = \boxed{28.1 \text{ ft}^3/\text{sec}}$$

$$\Delta h = 493.64 - 336.39$$

$$= \boxed{157.25 \text{ BTU/lbm} \quad [\text{decrease}]}$$

10. Although the ideal gas laws can be used, it is expedient to use the air tables.

At $460°\text{R} + 500°\text{R} = 960°\text{R}$,

$$h_1 = 231.06 \text{ BTU/lbm}$$
$$p_{r,1} = 10.610$$
$$\phi_1 = 0.7403$$

For isentropic compression,

$$p_{r,2} = 6p_{r,1}$$
$$= (6)(10.610) = 63.66$$

Searching the air table yields

$$T_2 = 1552°\text{R}$$
$$h_2 = 382.95 \text{ BTU/lbm}$$

The actual enthalpy is

$$h_2' = 231.06 + \frac{382.95 - 231.06}{0.65}$$
$$= 464.74 \text{ BTU/lbm}$$

This corresponds to $1855°\text{R}$, and $\phi_2' = 0.91129$.

$$W = \Delta h = 464.74 - 231.06$$
$$= 233.68 \text{ BTU/lbm}$$

$$\Delta s = 0.91129 - 0.7403 - \left(\frac{53.3}{778}\right)\ln(6)$$

$$= \boxed{0.04824 \text{ BTU/lbm-°R}}$$

Concentrates

1. First, assume isentropic compression and expansion. Refer to Fig. 7-14.

At a,

$$p_a = 100 \text{ psia}$$
$$h_a = 298.40 \text{ BTU/lbm}$$

At b,

$$p_b = 100 \text{ psia}$$
$$h_b = 1187.2 \text{ BTU/lbm}$$
$$s_b = 1.6026$$

At c,

$$p_c = 1 \text{ atm}$$
$$s_c = s_b = 1.6026$$
$$x_c = \frac{1.6026 - 0.3120}{1.4446} = 0.893$$
$$h_c = 180.07 + (0.893)(970.3)$$
$$= 1046.5 \text{ BTU/lbm}$$

At d, h and ν are essentially independent of pressure.

$$T_d = 80°\text{F} \quad [\text{subcooled}]$$
$$h_d = 48.02 \text{ BTU/lbm}$$
$$p_d = 14.7 \text{ psia}$$
$$v_d = 0.01608$$

At e,

$$p_e = p_a = 100 \text{ psia}$$
$$h_e = 48.02 + \frac{(0.01608)(100 - 14.7)(144)}{778}$$
$$= 48.27 \text{ BTU/lbm}$$

Due to the inefficiencies,

$$h_c' = 1187.2 - (0.80)(1187.2 - 1046.5)$$
$$= 1074.6 \text{ BTU/lbm}$$
$$h_e' = 48.02 + \frac{48.27 - 48.02}{0.6}$$
$$= 48.44 \text{ BTU/lbm}$$
$$\eta_{th} = \frac{(1187.2 - 1074.6) - (48.44 - 48.02)}{1187.2 - 48.44}$$
$$= \boxed{0.0985 \quad (9.85\%)}$$

2. Refer to Fig. 7.16.

At d,

$$p_d = 500 \text{ psia}$$
$$T_d = 1000°F$$
$$h_d = 1519.6 \text{ BTU/lbm}$$
$$s_d = 1.7363 \text{ BTU/lbm-°F}$$

At e,

$$p_e = 5 \text{ psia}$$
$$s_e = s_d = 1.7363$$
$$x_e = \frac{1.7363 - 0.2347}{1.6094} = 0.933$$
$$h_e = 130.13 + (0.933)(1001)$$
$$= 1064.1 \text{ BTU/lbm}$$

At f,

$$h_f = 130.13 \text{ BTU/lbm}$$

Because the turbine is 75% efficient,

$$h_e' = 1519.6 - (0.75)(1519.6 - 1064.1)$$
$$= 1178 \text{ BTU/lbm}$$

The mass flow rate is

$$\dot{m} = \frac{(200,000 \text{ kW})\left(1000 \frac{W}{kW}\right)\left(0.05692 \frac{BTU}{min\text{-}W}\right)}{\left(1519.6 \frac{BTU}{lbm} - 1178 \frac{BTU}{lbm}\right)\left(60 \frac{sec}{min}\right)}$$

$$= 555.4 \text{ lbm/sec}$$

$$Q_{out} = (555.4)(1178 - 130.13)$$

$$\boxed{= 5.82 \times 10^5 \text{ BTU/sec}}$$

3. Refer to Fig. 7.18.

At b,

$$p_b = 600 \text{ psia}$$
$$T_b = 486.21°F$$
$$h_b = 471.6 \text{ BTU/lbm}$$

At c,

$$h_c = 1203.2$$

At d,

$$p_d = 600 \text{ psia}$$
$$T_d = 600°F$$
$$h_d = 1289.9 \text{ BTU/lbm}$$

At e,

$$p_e = 200 \text{ psia}$$

From the Mollier diagram, assuming isentropic expansion,

$$h_e = 1187 \text{ BTU/lbm}$$
$$h_e' = 1289.9 - (0.88)(1289.9 - 1187)$$
$$= 1199.3 \text{ BTU/lbm}$$

At f,

$$p_f = 200 \text{ psia}$$
$$T_f = 600°F$$
$$h_f = 1322.1 \text{ BTU/lbm}$$
$$s_f = 1.6767$$

At g,

$$T_g = 60°F$$
$$s_g = s_f = 1.6767$$
$$x_g = \frac{1.6767 - 0.0555}{2.0393} = 0.795$$
$$h_g = 28.06 + (0.795)(1059.9) = 870.7 \text{ BTU/lbm}$$
$$h_g' = 1322.1 - (0.88)(1322.1 - 870.7)$$
$$= 924.9 \text{ BTU/lbm}$$

At h,

$$h_h = 28.06 \text{ BTU/lbm}$$
$$p_h = 0.2563 \text{ psia}$$
$$v_h = 0.01604 \text{ ft}^3/\text{lbm}$$

At a,

$$p_a = 600 \text{ psia}$$
$$h_a' = 28.06 + \frac{(0.01604)(600 - 0.2563)(144)}{(0.96)(778)}$$
$$= 29.9 \text{ BTU/lbm}$$

$$\eta_{th} = \frac{\begin{array}{c}(1289.9 - 29.9) + (1322.1 - 1199.3)\\ - (924.9 - 28.06)\end{array}}{(1289.9 - 29.9) + (1322.1 - 1199.3)}$$

$$\boxed{= 0.351 \ (35.1\%)}$$

4. Refer to p. 7-12 and the following diagram.

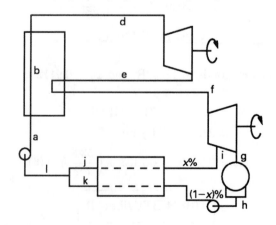

From problem 3,

$$h_b = 471.6 \text{ BTU/lbm}$$

$$h_d = 1289.9 \text{ BTU/lbm}$$

$$h_e' = 1199.3 \text{ BTU/lbm}$$

$$h_f = 1322.1 \text{ BTU/lbm}$$

$$h_g' = 924.9 \text{ BTU/lbm}$$

$$h_h = 28.04 \text{ BTU/lbm}$$

At I, the temperature is 270°F. Using the Mollier diagram and assuming isentropic expansion to 270°F,

$$h_i \approx 1170 \text{ BTU/lbm} \quad \text{[saturated]}$$

$$h_i' = 1322.1 - (0.88)(1322.1 - 1170)$$

$$= 1188.3 \text{ BTU/lbm}$$

At j the water is assumed to be saturated fluid at 270°F.

$$h_j = 238.84 \text{ BTU/lbm}$$

At k, the temperature is 270°F − 6°F = 264°F. Since the water is subcooled, enthalpy is a function of temperature only.

$$h_k = 232.83 \text{ BTU/lbm} \quad \begin{bmatrix} \text{includes condensate} \\ \text{pump work} \end{bmatrix}$$

From an energy balance in the heater,

$$(1-x)(h_k - h_h) = x(h_i' - h_j)$$

$$(1-x)(232.83 - 28.04) = x(1188.3 - 238.84)$$

$$204.79 = 1154.25x$$

$$x = 0.177$$

At l,

$$h_l = x(h_j) + (1-x)h_k$$

$$= (0.177)(238.84) + (1 - 0.177)(232.83)$$

$$= 233.89 \text{ BTU/lbm}$$

Since this is a subcooled liquid,

$$p_l = 38.5 \text{ psia}$$

$$v_l = 0.017132$$

$$T_l = 265°F$$

At a,

$$p_a = 600 \text{ psia}$$

$$h_a = 233.89 + \frac{(0.017132)(600 - 38.5)(144)}{(778)(0.96)}$$

$$= 235.7 \text{ BTU/lbm}$$

$$\eta_{th} = \frac{W_{out} - W_{in}}{Q_{in}}$$

$$= \frac{\begin{array}{c}(h_d - h_e') + (h_f - h_i') + (1-x)(h_i' - h_g') \\ -(h_a' - h_l)\end{array}}{(h_d - h_a) + (h_f - h_e')}$$

$$= \frac{\begin{array}{c}(1289.9 - 1199.3) + (1322.1 - 1188.3) + (1 - 0.177) \\ \times (1188.3 - 924.9) - (235.7 - 233.89)\end{array}}{(1289.9 - 235.7) + (1322.1 - 1199.3)}$$

$$= \boxed{0.373 \ (37.3\%)}$$

5. (a) Refer to Fig. 7.30. Use air tables to avoid having to evaluate k and c_v at the average temperatures.

At a,

$$V_a = 11 \text{ ft}^3$$

$$T_a = 460 + 80°F = 540°R$$

$$p_a = (14.2 \text{ psia})\left(144 \frac{\text{in}^2}{\text{ft}^2}\right) = 2044.8 \text{ lbf/ft}^2$$

$$m = \frac{pV}{RT} = \frac{\left(2044.8 \ \frac{\text{lbf}}{\text{ft}^2}\right)(11 \text{ ft}^3)}{\left(53.3 \ \frac{\text{ft-lbf}}{\text{lbm-°R}}\right)(540°R)}$$

$$= 0.781 \text{ lbm}$$

From air tables at 540°R,

$$v_{r,a} = 144.32$$

$$u_a = 92.04 \text{ BTU/lbm}$$

At b, the compression ratio is a ratio of volumes.

$$V_b = \left(\frac{1}{10}\right)V_a = \frac{11 \text{ ft}^3}{10}$$

$$= 1.1 \text{ ft}^3$$

Since the compression from a to b is isentropic,

$$v_{r,b} = \frac{v_{r,b}}{10} = \frac{144.32}{10}$$

$$= 14.432$$

Searching the air tables for this value of v_r,

$$T_b \approx 1314°R$$

$$u_b \approx 230.5 \text{ BTU/lbm}$$

At c, assume $T_c \approx 2300°R$. At $2300°R$, $u = 431.16$ BTU/lbm.

$$\bar{c}_v = \frac{\Delta u}{\Delta T} = \frac{431.16 \frac{BTU}{lbm} - 230.5 \frac{BTU}{lbm}}{2300°R - 1314°R}$$

$$= 0.2035 \text{ BTU/lbm-°R}$$

$$T_c = T_b + \frac{q_{in}}{\bar{c}_v} = T_b + \frac{Q_{in}}{m\bar{c}_v}$$

$$= 1314°R + \frac{160 \text{ BTU}}{(0.781 \text{ lbm})\left(0.2035 \frac{BTU}{lbm\text{-}°R}\right)}$$

$$= \boxed{2321°R}$$

From air tables,

$$v_{r,c} = 2.687$$
$$u_c = 435.67 \text{ BTU/lbm}$$

At d, since expansion is isentropic and the ratio of volumes is the same,

$$v_{r,d} = 10v_{r,c} = (10)(2.687)$$
$$= 26.87$$

Searching the air tables,

$$T_d = 1044°R$$
$$u_d = 180.38 \text{ BTU/lbm}$$

(b) The heat input is

$$q_{in} = \frac{160 \text{ BTU}}{0.781 \text{ lbm}} = 204.9 \text{ BTU/lbm}$$

The heat rejected during a constant volume process is

$$q_{out} = \Delta u$$

Heat is rejected between d and a. Therefore,

$$q_{out} = u_d - u_a = 180.38 \frac{BTU}{lbm} - 92.04 \frac{BTU}{lbm}$$

$$= 88.34 \text{ BTU/lbm}$$

The thermal efficiency is

$$\eta_{th} = \frac{q_{in} - q_{out}}{q_{in}} = \frac{204.9 \frac{BTU}{lbm} - 88.34 \frac{BTU}{lbm}}{204.9 \frac{BTU}{lbm}}$$

$$= \boxed{0.569 \ (56.9\%)}$$

6. *step 1:* Find the ideal mass of air ingested. The ideal mass of air taken in per second is

$$\dot{V}_i = \left(\begin{array}{c} \text{swept} \\ \text{volume} \end{array}\right)\left(\frac{\text{intake strokes}}{\text{sec}}\right)$$

The number of power strokes per second is

$$\frac{(2)(1200 \text{ rpm})(6 \text{ cylinders})}{\left(60 \frac{sec}{min}\right)(4 \text{ strokes})} = 60/\text{sec}$$

The swept volume is

$$V_s = \left(\frac{\pi}{4}\right)\left(\frac{4.25}{12}\right)^2\left(\frac{6}{12}\right)$$

$$= 0.04926 \text{ ft}^3$$

$$\dot{V}_1 = (60)(0.04926) = 2.956 \text{ ft}^3/\text{sec}$$

From $pV = mRT$, the ideal mass of the air in the swept volume is

$$m = \frac{(14.7)(144)(2.956)}{(53.3)(530)}$$

$$= 0.2215 \text{ lbm air/sec}$$

step 2: Find the CO_2 volume in the exhaust assuming complete combustion when the air/fuel ratio is 15.

Air is 76.85% nitrogen, so the nitrogen/fuel ratio is

$$N/F = (0.7685)(15) = 11.528$$

From $pV = mRT$, the nitrogen volume per pound of fuel burned is

$$V_N = \frac{mRT}{p} = \frac{(11.528)(55.2)(530°)}{(14.7)(144)}$$

$$= 159.3 \text{ ft}^3$$

The oxygen/fuel ratio is

$$O/F = (0.2315)(15) = 3.472$$

The oxygen volume per pound of fuel burned is

$$V_O = \frac{(3.472)(48.3)(530°)}{(14.7)(144)} = 41.99 \text{ ft}^3$$

When oxygen forms carbon dioxide, the chemical equation is

$$C + O_2 \longrightarrow CO_2$$

It takes one volume of oxygen to form one volume of CO_2. The percent of CO_2 in the exhaust is found from

$$\% \ CO_2 = \frac{\text{vol } CO_2}{\text{vol } CO_2 + \text{vol } O_2 + \text{vol } N_2}$$

$$\text{vol } N_2 = 159.3$$

$$\text{vol } CO_2 = x \quad [\text{unknown}]$$

$$\text{vol } O_2 = 41.99 - \binom{\text{oxygen used}}{\text{to make } CO_2}$$

$$= 41.99 - x$$

$$\frac{x}{x + 41.99 - x + 159.3} = 0.137$$

$$x = 27.58 \text{ ft}^3$$

Assuming complete combustion, the volume of CO_2 will be constant regardless of the amount of air used.

step 3: Calculate the excess air if the percent of CO_2 is 9%.

$$\% CO_2 = \frac{\text{vol } CO_2}{\text{vol } CO_2 + (\text{vol } O_2 - \text{vol } CO_2) + \text{vol } N_2 + \text{vol excess air}}$$

$$0.09 = \frac{27.58}{27.58 + (41.99 - 27.58) + 159.3 + \text{vol excess air}}$$

$$\text{vol excess air} = 105.2 \text{ ft}^3$$

From $pV = mRT$,

$$m_{\text{excess}} = \frac{(14.7)(144)(105.2)}{(53.3)(530°)}$$

$$= 7.883 \text{ lbm}$$

step 4: The actual air/fuel ratio is

$$15 + 7.883 = 22.883 \text{ lbm air/lbm fuel}$$

The actual air mass per sec is

$$\frac{\left(22.883 \frac{\text{lbm air}}{\text{lbm fuel}}\right)\left(28 \frac{\text{lbm fuel}}{\text{hr}}\right)}{3600 \frac{\text{sec}}{\text{hr}}} = 0.178 \text{ lbm/sec}$$

step 5:

$$\eta_v = \frac{0.178}{0.2215} = \boxed{0.804 \ (80.4\%)}$$

7. *step 1:*

altitude 1:   14.7 psia, $T = 60°F$
altitude 2: $z = 5000$ ft

step 2:

$$IHP_1 = \frac{1000 \text{ hp}}{0.80} = 1250 \text{ hp}$$

step 3: Not needed. (The problem says η is constant, not the friction horsepower.)

step 4:

$$\rho_1 = \frac{p}{RT} = \frac{(14.7)(144)}{(53.3)(520)}$$

$$= 0.0764 \text{ lbm/ft}^3$$

$$\rho_2 = 0.06592 \text{ at } 5000 \text{ ft}$$

step 5:

$$IHP_2 = (1250)\left(\frac{0.06592}{0.0764}\right) = 1078.5 \text{ hp}$$

step 6: $(0.80)(1078.5) = 862.8$

step 7: The original flow rate of fuel is

$$\dot{m}_{f,1} = (BHP_1)(BSFC_1) = (1000)(0.45)$$

$$= 450 \text{ lbm/hr}$$

step 8: $\dot{m}_{F,2} = \dot{m}_{F,1} = 450 \text{ lbm/hr}$

step 9: $BSFC_2 = \dfrac{450}{862.8} = \boxed{0.522}$

8. *Method 1:* Refer to Figure 7.36. As with other IC engines, the *compression ratio* is a ratio of volumes. First, assume isentropic operation.

At a,
$$p_a = 14.7 \text{ psia} = 2116.8 \text{ psfa}$$
$$T_a = 60°F = 520°R$$

For 1 lbm,

$$V_a = \frac{(1)(53.3)(520)}{2116.8} = 13.09 \text{ ft}^3$$

At b,

$$V_b = \frac{13.09}{5} = 2.618$$

$$T_b = (520)\left(\frac{13.09}{2.618}\right)^{1.4-1} = 989.9°R$$

$$p_b = \frac{(1)(53.3)(989.9)}{2.618} = 20{,}153 \text{ psfa}$$

At c,
$$T_c = 1500°F = 1960°R$$
$$p_c = 20{,}153 \text{ psfa}$$

At d:

$$p_d = 14.7 \text{ psia} = 2116.8 \text{ psfa}$$

$$T_d = (1960)\left(\frac{2116.8}{20{,}153}\right)^{\frac{1.4-1}{1.4}} = 1029.5$$

Now, include the inefficiencies.

$$T_a = 520°R$$

$$T_b' = 520°R + \frac{989.9 - 520}{0.83} = 1086°R$$

$$T_c = 1960°R$$

$$T_d' = 1960 - (0.92)(1960 - 1029.5) = 1103.9$$

$$\eta_{th} = \frac{(1960 - 1086) - (1102.9 - 520°R)}{1960 - 1086}$$

$$= \boxed{0.333 \ (33.3\%)}$$

Method 2:

At a,

$$p_a = 14.7 \text{ psia}$$

$$T_a = 520°R$$

$$v = 13.09 \text{ ft}^3/\text{lbm}$$

From the air table,

$$v_{r,a} = 158.58$$

$$p_{r,a} = 1.2147$$

$$h_a = 124.27 \text{ BTU/lbm}$$

At b,

$$v_{r,b} = \frac{v_{r,a}}{5} = \frac{158.58}{5}$$

$$= 31.716$$

$$T_b = 980°R$$

$$h_b = 236.02 \text{ BTU/lbm}$$

$$p_{r,b} = 11.430$$

$$p_b = \left(\frac{p_{r,b}}{p_{r,a}}\right) p_a = \left(\frac{11.430}{1.2147}\right)(14.7 \text{ psia})$$

$$= 138.3 \text{ psia}$$

At c,

$$T_c = 1960°R \quad \text{[given]}$$

$$p_c = p_b = 138.3 \text{ psia}$$

$$h_c = 493.64 \text{ BTU/lbm}$$

$$p_{r,c} = 160.37$$

At d,

$$p_d = 14.7 \text{ psia}$$

$$p_{r,d} = p_{r,c}\left(\frac{p_d}{p_c}\right) = (160.37)\left(\frac{14.7 \text{ psia}}{138.3 \text{ psia}}\right)$$

$$= 17.046$$

$$T_d = 1094°R$$

$$h_d = 264.49 \text{ BTU/lbm}$$

$$h_b' = 124.27 + \frac{236.02 - 124.27}{0.83}$$

$$= 258.9 \text{ BTU/lbm}$$

$$h_d' = 493.64 - (0.92)(493.64 - 264.49)$$

$$= 282.8 \text{ BTU/lbm}$$

$$\eta = \frac{493.64 - 258.9 - 282.8 + 124.27}{493.64 - 258.9}$$

$$= \boxed{0.325 \ (32.5\%)}$$

9. Refer to Fig. 7.38. From problem 8,

$$T_a = 520°R$$

$$T_b' = 1086°$$

$$T_d = 1960°$$

$$T_e' = 1103.9°$$

Assuming an ideal gas,

$$0.65 = \frac{T_c - T_b'}{T_e' - T_b'} = \frac{T_c - 1086}{1103.9 - 1086}$$

$$T_c = 1097.6$$

Assuming an ideal gas,

$$\eta_{th} = \frac{(1960 - 1103.9) - (1086 - 520)}{1960 - 1097.6}$$

$$= \boxed{0.336 \ (33.6\%)}$$

10. Shown below is one of N layers. Each layer consists or 24 tubes, only 3 of which are shown.

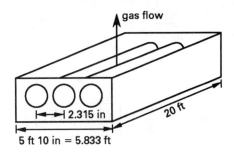

Approach: $q = UA\Delta T_m$

The log mean temperature difference (assuming counter flow operation) is

635°F ———— gases ———→ 470°F
$\Delta T = 350°F$ $\Delta T = 350°F$ 258°F
285°F ←———— water ———— 212°F

$$\Delta T_m = \frac{350 - 258}{\ln\left(\frac{350}{258}\right)} = 301.7°F$$

CHAPTER 7   POWER CYCLES

59

Since no information is given about the stack gases, assume that they consist of primarily nitrogen. The average gas temperature is

$$\left(\tfrac{1}{2}\right)(635+470)=552.5°F=1012.5°R$$

$$\bar{c}_{p,\text{nitrogen}} \approx 0.255 \text{ BTU/lbm-°F}$$

$$q = \dot{m}c_p\Delta T$$

$$= \left(191{,}000 \ \frac{\text{lbm}}{\text{hr}}\right)\left(0.255 \ \frac{\text{BTU}}{\text{lbm-°F}}\right)$$

$$\times (635-470)$$

$$= 8.036 \times 10^6 \text{ BTU/hr}$$

Since not enough information is given, h_i and h_o cannot be evaluated. U must be assumed. Assume $U_o = 10$ BTU/hr-ft²-°F.

$$A_o = \frac{q}{U_o\Delta T_m} = \frac{8.036\times 10^6}{(10)(301.7)}$$

$$= 2663.6 \text{ ft}^2$$

The tube area per bank is

$$24 \text{ tubes }(\pi)\left(\frac{1.315 \text{ ft}}{12}\right)(20 \text{ ft}) = 165.2 \text{ ft}^2$$

$$\text{no. of layers} = \frac{2663.6}{165.2} = \boxed{16.1 \quad [\text{say } 17]}$$

Answer will vary depending on value of U chosen.

Timed

1. (a) The number of power strokes per minute is

$$\frac{(2)(4600 \text{ rpm})(8 \text{ cylinders})}{4 \ \frac{\text{strokes}}{\text{cycle}}} = 18{,}400/\text{min}$$

The net work per cycle is

$$1500 - 1200 = 300 \text{ ft-lbf/cycle}$$

The indicated horsepower is

$$\text{IHP} = \frac{\left(\frac{18{,}400}{\text{min}}\right)(300 \text{ ft-lbf})}{33{,}000 \ \frac{\text{ft-lbf}}{\text{hp-min}}} = \boxed{167.27 \text{ hp}}$$

(b) The thermal efficiency is

$$\frac{300 \text{ ft-lbf}}{(1.27 \text{ BTU})\left(778 \ \frac{\text{ft-lbf}}{\text{BTU}}\right)} = \boxed{0.304 \ (30.4\%)}$$

Method 1:

(c) Assume an air fuel ratio of 15. Assume 14.7 psia and 70°F air.

The swept volume per cylinder is

$$\frac{265 \text{ in}^3}{(8)(12)^3} = 0.01917 \text{ ft}^3$$

The air density is

$$\rho = \frac{p}{RT} = \frac{(14.7)(144)}{(53.3)(460+70)}$$

$$= 0.07493 \text{ lbm/ft}^3$$

The air mass flow rate is. $\dot{V} = $ volumetric flow

$$\dot{m} = V N \rho = \dot{V}\rho$$

$$\dot{m}_A = \left[(0.01917 \text{ ft}^3)\left(\frac{18{,}400}{\text{min}}\right)\left(60 \ \frac{\text{min}}{\text{hr}}\right)\right]$$

$$\times \left(0.07493 \ \frac{\text{lbm}}{\text{ft}^3}\right) \qquad \dot{V} \Rightarrow \text{cfm}$$

$$= 1585.8 \text{ lbm/hr}$$

The fuel mass flow rate is

$$\dot{m}_F = \frac{\dot{m}_A}{R_{A/F}} = \frac{1585.8}{15}$$

$$= \boxed{105.7 \text{ lbm/hr}}$$

(d) The specific fuel comsumption is

$$\text{ISFC} = \frac{105.7 \ \frac{\text{lbm}}{\text{hr}}}{167.27 \text{ hp}} = \boxed{0.632 \text{ lbm/hp-hr}}$$

Method 2:

(c) Assume the heating value of gasoline is

$$\text{LHV} = 18{,}900 \text{ BTU/lbm}$$

Then, the fuel consumption is

$$\dot{m}_F = \frac{\left(1.27 \ \frac{\text{BTU}}{\text{cycle}}\right)\left(\frac{18{,}400}{\text{min}}\right)\left(60 \ \frac{\text{min}}{\text{hr}}\right)}{18{,}900 \ \frac{\text{BTU}}{\text{lbm}}}$$

$$= \boxed{74.18 \text{ lbm/hr}}$$

(d) $\text{ISFC} = \dfrac{74.18}{167.27} = \boxed{0.443 \text{ lbm/hp-hr}}$

2. Collect all enthalpies.

At 1: 1393.9 BTU/lbm
At 2: 1270 BTU/lbm
At 3: 1425.2 BTU/lbm
At 4: 1280 BTU/lbm
At 5: 1075 BTU/lbm
At 6: 69.73 BTU/lbm
At 7: $69.73 + 0.15 = 69.88$ BTU/lbm
At 8: 250.2 BTU/lbm
At 9: 253.1 BTU/lbm

(a) If the expansion had been isentropic to 200 psia,

$$h_2 = 1230 \text{ BTU/lbm} \quad \text{[from Mollier]}$$

$$\eta_{\text{isentropic}} = \frac{1393.9 - 1270}{1393.9 - 1230} = \boxed{0.756 \ (75.6\%)}$$

(b) Let x be the bleed fraction. From an energy balance in the heater,

$$h_8 = xh_4 + (1 - x)h_7$$

$$250.2 = x(1280) + (1 - x)(69.88)$$

$$x = 0.149$$

The thermal efficiency is

$$\eta_{\text{th}} = \frac{Q_{\text{in}} - Q_{\text{out}}}{Q_{\text{in}}}$$

$$= \frac{(h_1 - h_9) + (h_3 - h_2) - (1 - x)(h_5 - h_6)}{(h_1 - h_9) + (h_3 - h_2)}$$

$$= \frac{\begin{array}{c}(1393.9 - 253.1) + (1425.2 - 1270) \\ - (1 - 0.149)(1075 - 69.73)\end{array}}{(1393.9 - 253.1) + (1425.2 - 1270)}$$

$$= \boxed{0.34 \ (34\%)}$$

3. (a) Assume the isentropic efficiency is wanted. From air tables,

At 1,

$$T_1 = -10^\circ\text{F} = 450^\circ\text{R}$$

$$h_1 = 107.5 \text{ BTU/lbm}$$

$$p_{r,1} = 0.7329$$

$$p_1 = 8 \text{ psia}$$

At 2,

$$p_{r,2} = \left(\frac{p_2}{p_1}\right) p_{r,1} = \left(\frac{40}{8}\right)(0.7329)$$

$$= 3.6645$$

If the process was isentropic,

$$T_2 = 712^\circ\text{R}$$

$$h_2 = 170.47$$

However,

$$T_2' = 315^\circ\text{F} = 775^\circ\text{R}$$

$$h_2' = 185.75$$

$$\eta_{\text{isentropic}} = \frac{170.47 - 107.50}{185.73 - 107.50} = \boxed{0.805 \ (80.5\%)}$$

(b) For 35.7 psia, $T_{\text{sat}} \approx 20^\circ\text{F} = 480^\circ\text{R}$. For 172.4 psia, $T_{\text{sat}} \approx 120^\circ\text{F} = 580^\circ\text{R}$. For an ideal heat pump,

$$\text{COP} = \frac{580}{580 - 480} = \boxed{5.8}$$

(c) $$\text{COP} = \frac{(450 \text{ BTU})\left(1000 \ \dfrac{\text{W}}{\text{kW}}\right)}{(585 \text{ W})\left(3413 \ \dfrac{\text{BTU}}{\text{kW}}\right)} = \boxed{0.225}$$

(d) $$W_{\text{out}} = \frac{(600 \times 10^6 \text{ W})\left(3413 \ \dfrac{\text{BTU}}{\text{kW}}\right)}{1000 \ \dfrac{\text{W}}{\text{kW}}}$$

$$= 2.048 \times 10^9 \text{ BTU}$$

$$Q_{\text{in}} - Q_{\text{out}} = W_{\text{out}} - W_{\text{in}}$$

$$W_{\text{in}} \approx 0$$

$$Q_{\text{in}} = W_{\text{out}} + Q_{\text{out}}$$

$$= 2.048 \times 10^9 + 3.07 \times 10^9$$

$$= 5.118 \times 10^9 \text{ BTU}$$

$$\eta_{\text{th}} = \frac{Q_{\text{in}} - Q_{\text{out}}}{Q_{\text{in}}}$$

$$= \frac{(5.118 \times 10^9) - (3.07 \times 10^9)}{5.118 \times 10^9}$$

$$= \boxed{0.400}$$

(e) Assume a Carnot cycle.

$$\eta_{\text{th}} = \frac{(82^\circ + 460) - (40^\circ + 460)}{82^\circ + 460} = \boxed{0.0775}$$

4. If the horsepowers are the same,

(a)
$$HP_1 = HP_2$$

$$(\dot{m}_{F,1})(HV_1) = (\dot{m}_{F,2})(HV_2)$$

$$\dot{m}_F = (SFC)(HP)$$

$$(SFC_1)(HV_1) = (SFC_2)(HV_2)$$

$$\frac{SFC_2}{SFC_1} = \frac{HV_1}{HV_2} = \frac{23{,}200}{11{,}930}$$

$$= 1.945$$

$$\frac{SFC_2 - SFC_1}{SFC_1} = \frac{(1.945)(SFC_1) - SFC_1}{SFC_1}$$

$$= 0.945 \quad [94.5\% \text{ increase}]$$

(b)
$$\dot{m} = vA\rho$$

$$A = \frac{\dot{m}}{v\rho}$$

$$\dot{m}_2 = 1.945\,\dot{m}_1$$

$$\frac{A_2 - A_1}{A_1} = \frac{\dfrac{\dot{m}_2}{v\rho_2} - \dfrac{\dot{m}_1}{v\rho_1}}{\dfrac{\dot{m}_1}{v\rho_1}} = \frac{\dfrac{\dot{m}_2}{\rho_2} - \dfrac{\dot{m}_1}{\rho_1}}{\dfrac{\dot{m}_1}{\rho_1}}$$

$$= \frac{\dfrac{1.945}{\rho_2} - \dfrac{1}{\rho_1}}{\dfrac{1}{\rho_1}}$$

$$= (1.945)\left(\frac{\rho_1}{\rho_2}\right) - 1$$

Use interpolated special gravities.

At 68°F,

$$\frac{A_2 - A_1}{A_1} = (1.945)\left(\frac{0.724}{0.789}\right) - 1 = \boxed{0.785}$$

(c) Power is proportional to the weight flow and heating value.

$$\frac{P_2 - P_1}{P_1} = \frac{v_2 A_2 \rho_2 (HV)_2 - v_1 A_1 \rho_1 (HV)_1}{v_1 A_1 \rho_1 (HV)_1}$$

$$= \frac{\rho_2 (HV)_2 - \rho_1 (HV)_1}{\rho_1 (HV)_1}$$

$$= \frac{(0.789)(11{,}930) - (0.724)(23{,}200)}{(0.724)(23{,}200)}$$

$$= \boxed{-0.44 \quad (44\% \text{ decrease})}$$

5. (a) Work with 1 lbm. Assume ideal gases.

For the $3 \rightarrow 1$ process:

Find some composite property of the gas mixture.

$$p_1 = p_3 \left(\frac{T_1}{T_3}\right)^{\frac{k}{k-1}}$$

$$14.7 = (568.6)\left(\frac{520}{1600}\right)^{\frac{k}{k-1}}$$

$$0.02585 = (0.325)^{\frac{k}{k-1}}$$

$$\log(0.02585) = \left(\frac{k}{k-1}\right)\log(0.325)$$

$$\frac{k}{k-1} = 3.252$$

$$k = \boxed{1.444}$$

(b) The molar specific heat is

$$C_{p,\text{mixture}} = \frac{R^* k}{J(k-1)}$$

$$= \frac{\left(1545\,\dfrac{\text{ft-lbf}}{\text{lbmole-°F}}\right)(1.444)}{\left(778\,\dfrac{\text{ft-lbf}}{\text{BTU}}\right)(1.444 - 1)}$$

$$= 6.459\ \text{BTU/lbmole-°F}$$

For the constituent gases,

$$C_{p,\text{He}} = (\text{MW})(C_p)$$

$$= \left(4\,\frac{\text{lbm}}{\text{mole}}\right)\left(1.25\,\frac{\text{BTU}}{\text{lbm-°F}}\right)$$

$$= 5.0\ \text{BTU/lbmole-°F}$$

$$C_{p,\text{CO}_2} = (44)(0.205)$$

$$= 9.02\ \text{BTU/lbmole-°F}$$

Molar specific heat is volumetrically (mole fraction) weighted. Let

$$x = \frac{n_{\text{He}}}{n_{\text{He}} + n_{\text{CO}_2}} = \text{mole fraction He}$$

$$x C_{p,\text{He}} + (1 - x) C_{p,\text{CO}_2} = C_{p,\text{mixture}}$$

$$x\left(5.0\,\frac{\text{BTU}}{\text{lbmole-°F}}\right) + (1 - x)\left(9.02\,\frac{\text{BTU}}{\text{lbmole-°F}}\right)$$

$$= 6.459\ \text{BTU/lbmole-°F}$$

$$x = 0.637$$

Consider a 1-mole quantity of the gas mixture. The mass of He would be

$$m_{\text{He}} = x(\text{MW}) = (0.637\ \text{moles})\left(4\,\frac{\text{lbm}}{\text{mole}}\right)$$

$$= 2.548\ \text{lbm}$$

Similarly,

$$m_{CO_2} = (1 - 0.637 \text{ moles}) \left(44 \frac{\text{lbm}}{\text{mole}} \right)$$

$$= 15.972 \text{ lbm}$$

The molecular weight of the 1-mole of gas is

$$MW = m_{he} + m_{CO_2}$$

$$= 2.548 \text{ lbm} + 15.972 \text{ lbm}$$

$$= 18.52 \text{ lbm}$$

The gravimetric (mass) fraction of the gases are

$$G_{He} = \frac{m_{He}}{m_{He} + m_{CO_2}}$$

$$= \frac{2.548}{2.548 + 15.972}$$

$$= \boxed{0.138}$$

$$G_{CO_2} = 1 - G_{He} = 1 - 0.138$$

$$= \boxed{0.862}$$

$$C_v = C_p - \frac{R^*}{J}$$

$$= 6.458 \frac{\text{BTU}}{\text{lbmole-°F}} - \frac{1545 \frac{\text{ft-lbf}}{\text{lbmole-°F}}}{778 \frac{\text{ft-lbf}}{\text{BTU}}}$$

$$= 4.472 \text{ BTU/lbm-°F}$$

(C_v could also be found from x, $C_{v,He}$ and C_{v,CO_2}.)

$$W = c_v \Delta T = \left(\frac{C_v}{MW} \right) \Delta T$$

$$= \left(\frac{4.472 \frac{\text{BTU}}{\text{lbmole-°F}}}{18.52 \frac{\text{lbm}}{\text{lbmole}}} \right) (1600°R - 520°R)$$

$$= \boxed{260.8 \text{ BTU/lbm}}$$

(c)

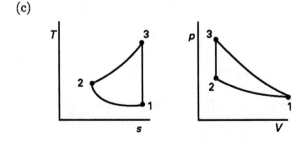

6. *step 2:* $IHP_1 = \frac{200}{0.86} = 232.6$

step 3: $FHP = 232.6 - 200 = 32.6$

step 4: $\rho_1 = \frac{p}{RT} = \frac{(14.7)(144)}{(53.3)(460 + 80)} = 0.0735$

$$\rho_2 = \frac{(12.2)(144)}{(53.3)(460 + 60)} = 0.0634$$

step 5: $IHP_2 = (232.6) \left(\frac{0.0634}{0.0735} \right) = 200.6$

step 6: $BHP_2 = 200.6 - 32.6 = \boxed{168.0 \text{ hp}}$

$$\eta = \frac{168}{200.6} = \boxed{0.837}$$

step 7: $\dot{m}_{F,1} = (0.48)(200) = 96 \text{ lbm/hr}$

$$\dot{m}_{A,1} = (22)(96) = 2112 \text{ lbm/hr}$$

$$\dot{V}_{A,1} = \frac{2112}{0.0735} = 28{,}735 \text{ ft}^3/\text{hr}$$

Since the engine speed is constant,

$$\dot{V}_{A,2} = \dot{V}_{A,1} = 28{,}735 \text{ ft}^3/\text{hr}$$

step 8: $\dot{m}_{A,2} = (28{,}735)(0.0634) = 1821.8 \text{ lbm/hr}$

Assume the metered fuel injection volume is the same.

$$\dot{m}_{F,2} = \dot{m}_{F,1} = 96 \text{ lbm/hr}$$

$$R_{A/F,2} = \frac{1821.8}{96} = \boxed{18.98}$$

$$BSFC_2 = \frac{96}{168} = \boxed{0.571 \text{ lbm/hp-hr}}$$

7. Work with the original system to find the steam flow.

$$h_1 = 1227.6 \text{ BTU/lbm} \quad \text{[superheated]}$$

$$h_2 = 38.04 \text{ BTU/lbm} \quad \text{[subcooled]}$$

$$h_3 = 147.92 \text{ BTU/lbm} \quad \text{[subcooled]}$$

Let x = fraction of steam in mixture.

$$147.92 = x(1227.6) + (1 - x)(38.04)$$

$$x = 0.0924$$

The steam flow is

$$(0.0924)(2000) = 184.8 \text{ lbm/hr}$$

Since the pressure drop across the heater is 5 psi,

$$p_4 = p_6 + 5 = 20 + 5 = 25 \text{ psia}$$

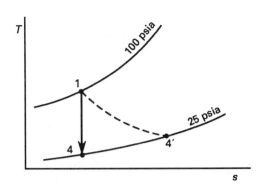

From the Mollier diagram,

$$h_4 \approx 1116 \text{ BTU/lbm} \quad \text{[liquid-vapor mixture]}$$
$$h_4' = 1227.6 - (0.60)(1227.6 - 1116)$$
$$= 1160.6 \text{ BTU/lbm}$$
$$W_{\text{out}} = (0.96)\left(184.8 \, \frac{\text{lbm}}{\text{hr}}\right)$$
$$\times \left(1227.6 \, \frac{\text{BTU}}{\text{lbm}} - 1160.6 \, \frac{\text{BTU}}{\text{lbm}}\right)$$
$$= \boxed{11{,}886 \text{ BTU/hr}}$$

Let x = fraction of steam in.

$$147.92 = x(1160.6) + (1 - x)(38.04)$$
$$x = 0.0979$$
$$\dot{m}_6 = \frac{184.8}{0.0979} = \boxed{1888 \text{ lbm/hr}}$$

8.

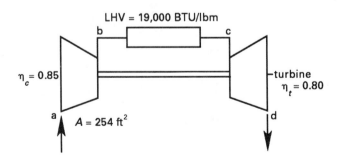

Initially,

$$Q = 50{,}000 \text{ ft}^3/\text{min}$$
$$BHP = 6000$$
$$BSCF = 0.609$$

At a,

$$p_a = 12 \text{ psia}$$
$$T_a = 35°\text{F} = 495°\text{R}$$
$$h_a = 118.28 \text{ BTU/lbm} \quad \text{[from air table]}$$
$$p_{r,a} = 1.0224$$
$$\text{air density} = \frac{p}{RT} = \frac{(12)(144)}{(53.3)(35 + 460)}$$
$$= 0.0655 \text{ lbm/ft}^3$$

The air mass flow rate is

$$\dot{m} = Q\rho = (50{,}000)(0.0655)$$
$$= 3275 \text{ lbm/min}$$

At b,

$$p_b = 8p_a = (8)(12)$$
$$= 96 \text{ psia}$$

Assuming isentropic compression,

$$p_{r,b} = 8p_{r,a} = (8)(1.0224)$$
$$= 8.1792$$

This $p_{r,b}$ corresponds to

$$T_b = 893°\text{R}$$
$$h_b = 214.54$$

But compression is not isentropic.

$$h_b' = 118.28 + \frac{214.54 - 118.28}{0.85}$$
$$= 231.53 \text{ BTU/lbm}$$
$$T_b' = 962°\text{R}$$
$$W_{\text{compression}} = 231.53 - 118.28 = 113.25 \text{ BTU/lbm}$$

At c,

$$T_c = 1800°\text{F} = 2260°\text{R} \quad \text{[no change if moved]}$$
$$p_c = 96 \text{ psia} \quad \begin{bmatrix} \text{in reality, some pressure} \\ \text{drop would be expected} \\ \text{across combustor} \end{bmatrix}$$
$$h_c = 577.51 \text{ BTU/lbm}$$
$$p_{r,c} = 286.6$$

The energy requirements are

$$\dot{m}\Delta h = (3275)(577.51 - 231.53)$$
$$= 1.133 \times 10^6 \text{ BTU/min}$$

The ideal fuel rate is

$$\frac{\left(1.133 \times 10^6 \, \dfrac{\text{BTU}}{\text{min}}\right)\left(60 \, \dfrac{\text{min}}{\text{hr}}\right)}{19{,}000 \, \dfrac{\text{BTU}}{\text{lbm}}} = 3577.9 \text{ lbm/hr}$$

The ideal BSFC is

$$\frac{3577.9}{6000} = 0.596 \text{ lbm/hp-hr}$$

The actual BSFC is given. The combustor efficiency is

$$\frac{0.596}{0.609} = 0.979 \quad (97.9\%)$$

At d,

$$p_d = 12 \text{ psia} \quad \left[\begin{array}{c} \text{determined by} \\ \text{atmospheric conditions} \end{array}\right]$$

If isentropic expansion,

$$p_{r,d} = \frac{286.6}{8} = 35.825$$

$$T_d = 1335°\text{R}$$

$$h_d = 325.99 \text{ BTU/lbm}$$

$$h'_d = 577.51 - (0.80)(577.51 - 325.99)$$

$$= 376.29 \text{ BTU/lbm}$$

$$W_{\text{turbine}} = 577.51 - 376.29 = 201.22$$

The theoretical net horsepower, corrected for non-isentropic expansion and subtracting work of compression, is

$$IHP = \dot{m}\Delta h J$$

$$= \frac{\left(3275 \frac{\text{lbm}}{\text{min}}\right)\left(201.22 \frac{\text{BTU}}{\text{lbm}} - 113.25 \frac{\text{BTU}}{\text{lbm}}\right)}{33,000 \frac{\text{ft-lbf}}{\text{hp-min}}} \times \left(778 \frac{\text{ft-lbf}}{\text{BTU}}\right)$$

$$= 6792 \text{ hp}$$

The friction horsepower is

$$FHP = IHP - BHP = 6792 - 6000$$

$$= 792 \text{ hp}$$

At 0 altitude:

At a,

$$p = 14.7 \text{ psia}$$

$$T = 70°\text{F} = 530°\text{R}$$

$$h_a = 126.6 \text{ BTU/lbm}$$

$$p_{r,a} = 1.2983$$

$$\rho = \frac{(14.7)(144)}{(53.3)(530)} = 0.0749 \text{ lbm/ft}^3$$

$$\dot{m}_{\text{air}} = (50,000)(0.0749) = 3745 \text{ lbm/min}$$

At b,

$$p_b = (8)(14.7) = 117.6$$

$$p_{r,b} = (8)(1.2983) = 10.3864$$

$$T_b = 954°\text{R}$$

$$h_b = 229.57 \text{ BTU/lbm}$$

$$h'_b = 126.6 + \frac{229.57 - 126.6}{0.85}$$

$$= 247.74 \text{ BTU/lbm}$$

$$T'_b = 1027°\text{R}$$

$$W_{\text{compression}} = 247.74 - 126.6 = 121.14 \text{ BTU/lbm}$$

At c,

$$p_c = 117.6 \quad \left[\begin{array}{c} \text{assuming no pressure drop} \\ \text{across the combustor} \end{array}\right]$$

$$T_c = 2260°\text{R}$$

$$h_c = 577.51 \text{ BTU/lbm}$$

$$p_{r,c} = 286.6$$

Energy requirements are

$$\dot{m}\Delta h = (3745)(577.51 - 247.74)$$

$$= 1.235 \times 10^6 \text{ BTU/min}$$

Assuming a constant combustion efficiency of 97.9%,

$$\text{fuel rate} = \frac{(1.235 \times 10^6)(60)}{(19,000)(0.979)} = 3983.7 \text{ lbm/hr}$$

At d,

$$p_d = 14.7 \text{ psi}$$

$$p_{r,d} = \frac{286.6}{8} = 35.825 \quad [\text{no change}]$$

$$T_d = 1335°\text{R}$$

$$h_d = 325.99 \text{ BTU/lbm}$$

$$h'_d = 376.29 \text{ BTU/lbm} \quad [\text{no change}]$$

$$W'_{\text{turbine}} = 201.22 \quad [\text{no change}]$$

The horsepower corrected for nonisentropic expansion is

$$IHP = \frac{(3745)(201.22 - 121.14)(778)}{33,000}$$

$$= 7070 \text{ hp}$$

Assuming frictional horsepower is constant,

$$BHP = IHP - FHP$$

$$= 7070 - 792$$

$$= \boxed{6278 \text{ hp}}$$

$$BSFC = \frac{3983.7}{6278}$$

$$= \boxed{0.635 \text{ lbm/hp-hr}}$$

COMPRESSIBLE FLUID DYNAMICS

Warmups

1.
$$T_1 = 460 + 150°F$$
$$= 610°R$$

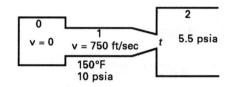

The velocity of sound at point 1 is

$$c_1 = \sqrt{kg_cRT} = \sqrt{(1.4)(32.2)(53.3)(610)}$$
$$= 1210.7 \text{ ft/sec}$$

The entrance Mach number is

$$M_1 = \frac{v_1}{c_1} = \frac{750}{1210.7}$$

$$= \boxed{0.619 \quad [\text{say } 0.62]}$$

From compressible flow tables (CFT),

$$\frac{T_1}{T_0} = \frac{T}{TT} = 0.9286$$

$$\frac{p_1}{p_0} = \frac{p}{Tp} = 0.7716$$

$$T_0 = \frac{610}{0.9286} = 656.9°R$$

$$p_0 = \frac{10}{0.7716} = 12.96 \text{ psia}$$

$$\frac{p_2}{p_0} = \frac{5.5}{12.96} = 0.424$$

The critical pressure ratio for air is 0.5283. Since $p_2/p_0 < 0.5283$, the flow is choked, and $M_t = 1$. For $M = 1$,

$$\frac{p}{p_0} = \frac{p}{TP} = 0.5283$$

$$\frac{T}{T_0} = \frac{T}{TT} = 0.8333$$

$$p = (0.5283)(12.96) = \boxed{6.85 \text{ psia}}$$

$$T = (0.8333)(656.9) = \boxed{547.4°R}$$

2. Assume STP is 14.7 psia and 32°F.

$$T_x = 460 + 32 = 492°R$$
$$p_x = 14.7 \text{ psia}$$
$$c = \sqrt{(1.4)(32.2)(53.3)(492)} = 1087.3 \text{ ft/sec}$$
$$M = \frac{2000}{1087.3} = 1.84$$

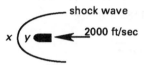

Assume that the bullet is stationary and the STP air is moving at $M = 1.84$. From the compressible flow tables (CFT) for $M = 1.84$,

$$\frac{p_x}{TP_y} = 0.2060$$

$$\frac{T_x}{TT_x} = 0.5963$$

$$TP_y = \frac{14.7}{0.2060} = 71.36 \text{ psia}$$

Since a shock wave is adiabatic,

$$TT_x = TT_y = \frac{492}{0.5963} = 825.1$$

From air tables,

$$h \approx \boxed{197.9 \text{ BTU/lbm}}$$

3. Assume the pressure and temperature given are the chamber properties.

$$TT = 240°F = 700°R$$
$$TP = 160 \text{ psia}$$

$20/160 < 0.5283$, so the nozzle is supersonic and the throat flow is sonic.

The total density is

$$TD = \frac{TP}{(R)(TT)} = \frac{(160)(144)}{(53.3)(700)}$$
$$= 0.6175 \text{ lbm/ft}^3$$

At the throat,

$$M = \boxed{1}$$

$$\frac{D}{TD} = 0.6339$$

$$\frac{T}{TT} = 0.8333$$

$$T^* = (700)(0.8333) = 583.3°R$$

$$D^* = (0.6175)(0.6339) = 0.3914 \text{ lbm/ft}^3$$

$$c^* = \sqrt{(1.4)(53.3)(32.2)(583.3)}$$

$$= 1183.9 \text{ ft/sec}$$

$$A^* = \frac{\dot{m}}{D^* c^*}$$

$$= \frac{4.5 \dfrac{\text{lbm}}{\text{sec}}}{\left(0.3914 \dfrac{\text{lbm}}{\text{ft}^3}\right)\left(1183.9 \dfrac{\text{ft}}{\text{sec}}\right)}$$

$$= \boxed{0.00971 \text{ ft}^2}$$

At the exit,

$$\frac{p}{TP} = \frac{20}{160} = 0.125$$

Searching the CFT gives

$$M \approx \boxed{2.01}$$

At $M = 2.01$,

$$\frac{A}{A^*} = 1.7016$$

$$A_e = (1.7016)(0.00971)$$

$$= \boxed{0.01652 \text{ ft}^2}$$

4.
$$c = \sqrt{(1.4)(32.2)(53.3)(460 + 60)}$$

$$= 1117.8 \text{ ft/sec}$$

$$M = \frac{2700}{1117.8} = 2.415$$

From the appropriate table or figure,

$$\theta_{max} \approx \boxed{45°}$$

The actual value may be smaller.

5.
$$TT = 460 + 80 = 540°R$$

$$TP = 100 \text{ psia}$$

From the CFT at $M = 2$,

$$\frac{p}{TP} = 0.1278$$

$$\frac{T}{TT} = 0.5556$$

$$\frac{A}{A^*} = 1.6875$$

$$T = (0.5556)(540) = \boxed{300°R}$$

$$A = (1.6875)(1) = \boxed{1.6875 \text{ ft}^2}$$

$$p = (0.1278)(100) = 12.78 \text{ psia}$$

$$c = \sqrt{(1.4)(32.2)(53.3)(300)}$$

$$= 849 \text{ ft/sec}$$

$$v = Mc = (2)(849)$$

$$= 1698 \text{ ft/sec}$$

$$D = \rho = \frac{p}{RT}$$

$$= \frac{(12.78)(144)}{(53.3)(300)}$$

$$= 0.1151 \text{ lbm/ft}^3$$

$$\dot{m} = vA\rho = (1698)\left(\frac{1.6875}{144}\right)(0.1151)$$

$$= \boxed{2.29 \text{ lbm/sec}}$$

6. From the normal shock tables at $M_x = 2$,

$$M_y = 0.5744$$

$$\frac{T_y}{T_x} = 1.687$$

$$T_y = (1.687)(500) = 843.5°R$$

$$c_y = \sqrt{(1.4)(32.2)(53.3)(843.5)} = 1423.6 \text{ ft/sec}$$

$$v_y = Mc = (0.5744)(1423.6)$$

$$= \boxed{818 \text{ ft/sec}}$$

7.
$$TP = 100 \text{ psia}$$
$$TT = 460 + 70 = 530°\text{R}$$

Searching the CFT for $A/A^* = 1.555$,

$$M = \boxed{1.9}$$

$$\frac{T}{TT} = 0.5807$$

$$T = (0.5807)(530) = \boxed{307.8°\text{R}}$$

$$\frac{p}{TP} = 0.1492$$

$$p = (0.1492)(100) = \boxed{14.92 \text{ psia}}$$

8. Since v is unknown, assume that the static temperature is $40 + 460 = 500°\text{R}$. Then,

$$\rho = \frac{p}{RT} = \frac{(10)(144)}{(53.3)(500)}$$
$$= 0.054 \text{ lbm/ft}^3$$

$$v = \frac{\dot{m}}{A\rho} = \frac{20 \frac{\text{lbm}}{\text{sec}}}{(1 \text{ ft}^2)\left(0.054 \frac{\text{lbm}}{\text{ft}^3}\right)}$$
$$= 370.4 \text{ ft/sec}$$

At 500°F,

$$c = \sqrt{(1.4)(32.2)(53.3)(500)} = 1096.1 \text{ ft/sec}$$
$$M = \frac{370.4}{1096.1} = 0.338 \quad [\text{say } 0.34]$$

At $M = 0.34$,
$$\frac{T}{TT} = 0.9774$$

A closer approximation to T is

$$T = (0.9774)(500) = 488.7°\text{R}$$
$$\rho = \frac{p}{RT} = \frac{(10)(144)}{(53.3)(488.7)}$$
$$= 0.0553 \text{ lbm/ft}^3$$
$$v = \frac{\dot{m}}{A\rho} = \frac{20}{(1)(0.0553)}$$
$$= 361.7 \text{ ft/sec}$$
$$c = \sqrt{(1.4)(32.2)(53.3)(488.7)} = 1083.6 \text{ ft/sec}$$
$$M = \frac{361.7}{1083.6} = 0.334 \quad [\text{say } 0.33]$$

At $M = 0.33$,

$$\frac{A}{A^*} = 1.8707$$
$$A_{\text{smallest}} = \frac{1}{1.8707}$$
$$= \boxed{0.535}$$

9.
$$\frac{p_x}{TP_y} = \frac{1.38}{20} = 0.069$$

Searching the CFT, $M = \boxed{3.3}$

10.
$$h_1 = 1428.9 \text{ BTU/lbm}$$

Using the Mollier diagram and assuming isentropic expansion,

$$h_2 = 1362 \text{ BTU/lbm}$$

Assuming $v_1 = 0$,

$$v = \sqrt{(2)(32.2)(778)(1428.9 - 1362)}$$
$$= \boxed{1830.8 \text{ ft/sec}}$$

Concentrates

1. At that point,

$$c = \sqrt{(1.4)(32.2)(53.3)(1000)} = 1550.1 \text{ ft/sec}$$
$$M = \frac{600}{1550.1} = \boxed{0.387}$$

At $M = 0.39$,

$$\frac{T}{TT} = 0.9705$$
$$\frac{p}{TP} = 0.9004$$
$$\frac{A}{A^*} = 1.6243$$
$$TT = \frac{1000}{0.9705} = 1030.4$$
$$TP = \frac{50}{0.9004} = 55.5$$
$$A^* = \frac{0.1}{1.6234} = \boxed{0.0616}$$

At $M = 1$,

$$\frac{p}{TP} = 0.5283$$

$$\frac{T}{TT} = 0.8333$$

$$p^* = (55.5)(0.5283) = \boxed{29.32 \text{ psia}}$$

$$T^* = (1030.4)(0.8333) = 858.6°\text{R}$$

2. *Method 1:* For $p_1 = 200$ psia and $T_1 = 600°$F steam, read $h_1 = 1322.1$ BTU/lbm and $v_1 = 3.060$ ft^3/lbm. From the superheat table, locate this point on the Mollier diagrams.

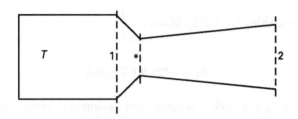

At $600°$F $= 1060°$R, $k_1 \approx 1.29$ and $c_p \approx 0.473$ BTU/lbm-°R. The speed of sound at 1 is

$$c_1 = \sqrt{kg_cRT} = \sqrt{(1.29)(32.2)(85.8)(1060)}$$

$$= 1943.7 \text{ ft/sec}$$

$$M_1 = \frac{v_1}{c_1} = \frac{300 \dfrac{\text{ft}}{\text{sec}}}{1943.7 \dfrac{\text{ft}}{\text{sec}}}$$

$$= 0.154 \quad [\text{use } 0.15]$$

From the CFT for $k = 1.30$ and $M_1 = 0.15$,

$$\frac{p}{TP} = 0.9855$$

$$\frac{T}{TT} = 0.9966$$

Therefore, the total properties are

$$TP = \frac{200 \text{ psia}}{0.9855} = 202.9 \text{ psia}$$

$$TT = \frac{1060°\text{R}}{0.9966} = 1063.6°\text{R}$$

The original total enthalpy can be found from

$$v_1 = \sqrt{2g_cJ(h_T - h_1) + v_0^2}$$

$$300 \text{ ft/sec} = \sqrt{\begin{array}{c} (2)\left(32.2 \dfrac{\text{ft-lbm}}{\text{lbf-sec}^2}\right)\left(778 \dfrac{\text{ft-lbf}}{\text{BTU}}\right) \\ \times \left(h_T - 1322.1 \dfrac{\text{BTU}}{\text{lbm}}\right) + (0)^2 \end{array}}$$

$$h_T = 1323.9 \text{ BTU/lbm}$$

Alternatively:

$$h_T \approx h_1 + c_p(TT - T_1)$$

$$= 1322.1 \frac{\text{BTU}}{\text{lbm}}$$

$$+ \left(0.473 \frac{\text{BTU}}{\text{lbm-°R}}\right)(1063°\text{R} - 1060°\text{R})$$

$$= 1323.8 \text{ BTU/lbm}$$

$$p^* \approx (0.57)TP = (0.57)(202.9 \text{ psia})$$

$$= 115.7 \text{ psia}$$

From the Mollier diagram, assuming isentropic expansion from $p_1 = 200$ psia to 115.7 psia,

$$h^* \approx 1263 \text{ BTU/lbm}$$

$$T^* \approx 470°\text{F} = 930°\text{R}$$

$$v^* = \sqrt{2g_cJ(h_1 - h^*) + v_1}$$

$$= \sqrt{\begin{array}{c} (2)\left(32.2 \dfrac{\text{ft-lbm}}{\text{lbf-sec}^2}\right)\left(778 \dfrac{\text{ft-lbf}}{\text{BTU}}\right) \\ \times \left(1322.1 \dfrac{\text{BTU}}{\text{lbm}} - 1263 \dfrac{\text{BTU}}{\text{lbm}}\right) + \left(300 \dfrac{\text{ft}}{\text{sec}}\right)^2 \end{array}}$$

$$= 1746 \text{ ft/sec}$$

$$\rho^* = \frac{p^*}{RT^*} = \frac{(115.7 \text{ psia})\left(144 \dfrac{\text{in}^2}{\text{ft}^2}\right)}{(85.8)(930°\text{R})}$$

$$= 0.209 \text{ lbm/ft}^3$$

$$A^* = \frac{\dot{m}}{\rho^* v^*}$$

$$= \frac{3 \dfrac{\text{lbm}}{\text{sec}}}{\left(0.209 \dfrac{\text{lbm}}{\text{ft}^3}\right)\left(1746 \dfrac{\text{ft}}{\text{sec}}\right)}$$

$$= \boxed{0.00822 \text{ ft}^2}$$

Method 2:

$$h_1 = 1322.1$$

From the Mollier diagram, assuming isentropic expansion,

$$h_2 = 1228 \text{ BTU/lbm}$$

$$h_2' = 1322.1 - (0.85)(1322.1 - 1228)$$

$$= 1242.1 \text{ BTU/lbm}$$

Knowing $h = 1242.1$ BTU/lbm and $p = 80$ psi establishes (from detailed superheat tables)

$$T_2' = 420°F$$

$$v_2' = 6.383$$

$$p_2 = \frac{1}{v_2} = \frac{1}{6.383}$$

$$= 0.1567 \text{ lbm/ft}^3$$

$$v_2' = \sqrt{(2)(32.2)(778)(1322.1 - 1242.1) + (300)^2}$$

$$= 2024.4 \text{ ft/sec at exit}$$

$$A_e = \frac{\dot{m}}{v\rho} = \frac{3 \dfrac{\text{lbm}}{\text{sec}}}{\left(2024.4 \dfrac{\text{ft}}{\text{sec}}\right)\left(0.1567 \dfrac{\text{lbm}}{\text{ft}^3}\right)}$$

$$= 0.009457$$

For $420°F + 460 = 880°R$ steam,

$$k_{\text{steam}} = 1.31$$

$$R_{\text{steam}} = 85.8$$

$$c = \sqrt{(1.31)(32.2)(85.8)(880)}$$

$$= 1784.6 \text{ ft/sec}$$

$$M = \frac{2024.4}{1784.6} = 1.13$$

From the CFT at $M = 1.13$ and $k = 1.30$,

$$\frac{A}{A^*} = 1.0139$$

$$A^* = \frac{A_e}{1.0139} = \frac{0.009457}{1.0139}$$

$$= \boxed{0.00933 \text{ ft}^2}$$

Timed

1. This is a Fanno flow problem. Check for choked flow.

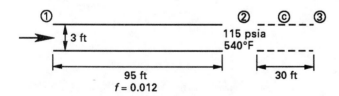

Since p is decreasing, the flow is initially subsonic.

Method 1: At point 2, from the superheat tables,

$$v_2 = 5.066 \text{ ft}^3/\text{lbm}$$

$$\text{v} = \frac{\dot{m}}{A\rho} = \frac{\dot{m}v}{A}$$

$$= \frac{(35,200)(5.066)}{(3600)\left(\dfrac{\pi}{4}\right)\left(\dfrac{3}{12}\right)^2}$$

$$= 1009 \text{ ft/sec}$$

Assume $k = 1.29$ at $1000°R$.

$$c = \sqrt{kg_cRT}$$

$$= \sqrt{(1.29)(32.2)(85.8)(540 + 460)}$$

$$= 1888 \text{ ft/sec}$$

The Mach number at point 2 is

$$M_2 = \frac{1009}{1888} = 0.534$$

The distance from point 2 to where the flow becomes choked is

$$x_{\max} = \left[\frac{\dfrac{3}{12}}{(4)(0.012)}\right]\left[\frac{1 - (0.534)^2}{(1.29)(0.534)^2} + \frac{1.29 + 1}{(2)(1.29)}\right.$$

$$\left. \times \ln\left(\frac{(1.29 + 1)(0.534)^2}{(2)\left[1 + \left(\dfrac{1.29 - 1}{2}\right)(0.534)^2\right]}\right)\right]$$

$$= 4.76 \text{ ft}$$

$$\boxed{\text{The flow will be choked in less than 30 ft.}}$$

Method 2: Use the Fanno flow table for $k = 1.3$.

At $M_2 = 0.53$,
$$\frac{4fL_{\max}}{D} = 0.949$$

At $M_c = 1$,

$$\frac{4fL_{\max}}{D} = 0$$

$$L_{\max} = \frac{(0.949 - 0)\left(\dfrac{3}{12}\right)}{(4)(0.012)}$$

$$= 4.94 \text{ ft}$$

2. Since the pressure drops to 14.7 psia, this must be a supersonic nozzle.

Assume that the given properties are total properties. At the throat,

$$M = 1$$

$$\frac{T}{TT} = 0.8333$$

$$\frac{p}{TP} = 0.5283$$

$$T_{\text{throat}} = (0.8333)(660) = 550°\text{R}$$

$$p_{\text{throat}} = (0.5283)(160) = 84.53 \text{ psia}$$

$$c^* = \sqrt{kg_cRT} = \sqrt{(1.4)(32.2)(53.3)(550)}$$
$$= 1150 \text{ ft/sec}$$

Since air is an ideal gas,

$$\rho_{\text{throat}} = \frac{p}{RT} = \frac{(84.53)(144)}{(53.3)(550)}$$
$$= 0.415 \text{ lbm/ft}^3$$

The overall nozzle efficiency is $C_D = 0.90$. From $\dot{m} = C_D A v \rho_1$,

$$A^* = \frac{\dot{m}}{C_D v \rho} = \frac{3600}{(3600)(1150)(0.415)(0.90)}$$
$$= \boxed{0.002328 \text{ ft}^2}$$

$$D_{\text{throat}} = \sqrt{\frac{4A}{\pi}} = \sqrt{\frac{(4)(0.002328)}{\pi}}$$
$$= 0.0544 \text{ ft}$$

At the exit,
$$p = 14.7$$

$$\frac{p}{TP} = \frac{14.7}{160} = 0.0919$$

Finding this value in the isentropic flow tables yields

$$M = 2.22$$

$$\frac{A}{A^*} = 2.041$$

$$A_{\text{exit}} = (2.041)(0.002328) = 0.004751 \text{ ft}^2$$

$$D_{\text{exit}} = \sqrt{\frac{(4)(0.004751)}{\pi}}$$
$$= 0.0778 \text{ ft}$$

The longitudinal distance from the throat to the exit is

$$x = \frac{0.0778 - 0.0544}{(2)(\tan 3°)} = 0.223 \text{ ft}$$

The entrance velocity is not known, so the entrance area cannot be found. However, the longitudinal distance from the entrance to the throat is

$$(0.05)(0.223) = 0.0112 \text{ ft}$$

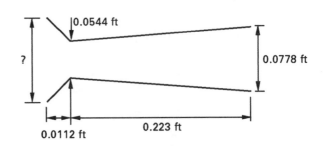

COMBUSTION

1.

	G	AW	G/AW
C	40.0	12	3.33
H	6.7	1	6.7
O	53.3	16	3.33

$\longrightarrow$ $\boxed{CH_2O}$

2. One mole of any gas occupies a volume of $22.4\,\ell$.

$$\frac{200}{22.4} = 8.929 \text{ gmol of } CH_4$$

One mole of methane has a molecular weight of 16 g, so the methane has a mass of

$$(16)(8.929) = 142.86 \text{ g}$$

$$(142.86 \text{ g})\left(0.0022046\,\frac{lbm}{g}\right) = 0.315 \text{ lbm}$$

The available energy is

$$(0.315 \text{ lbm})\left(24{,}000\,\frac{BTU}{lbm}\right) = 7560 \text{ BTU}$$

$$\Delta T = (95 - 15)\left(\frac{9}{5}\right) = 144°F$$

$$q = \frac{mc_p\Delta T}{\eta}$$

$$m = \frac{(0.5)(7560)\left(0.4536\,\frac{kg}{lbm}\right)}{(1)(144)}$$

$$= \boxed{11.91 \text{ kg}}$$

3.

$$C_3H_8 + 5O_2 \longrightarrow 3CO_2 + 4H_2O$$
$$(MW) \quad 44 + 160 \longrightarrow 132 + 72$$

$$\frac{CO_2}{C_3H_8} = \frac{132}{44} = \frac{x}{15}$$

$$x = 45 \text{ lbm/hr}$$

$$V = \frac{mRT}{p} = \frac{(45)(35.1)(530°R)}{(14.7)(144)}$$

$$= \boxed{395.5 \text{ ft}^3/\text{hr}}$$

4. For every volume of oxygen in the air, there are approximately 3.78 volumes of nitrogen. The stoichiometric reaction equation is

$$CH_4 + 2O_2 + (2)(3.78)N_2$$
$$\text{volumes} \quad 1 \qquad 2 \qquad\quad 7.56$$

$$\longrightarrow CO_2 + 2H_2O + (2)(3.78)N_2$$
$$\qquad\quad 1 \qquad 2 \qquad\quad 7.56$$

The stoichiometric nitrogen volume (with the excess) is

$$V_{N_2} = (1.3)\left(7.56\,\frac{ft^3\,N_2}{ft^3\,CH_4}\right)(4000\,ft^3\,N_2)$$

$$= 39{,}312\,ft^3\,N_2$$

Air is 79.1% nitrogen by volume, so by Dalton's Law, the partial pressure of the nitrogen is

$$p_{N_2} = (0.791)(15 \text{ psia}) = 11.865 \text{ psia}$$

The mass of nitrogen is

$$m_{N_2} = \frac{pV}{RT} = \frac{\left(11.865\,\frac{lbf}{in^2}\right)\left(144\,\frac{in^2}{ft^2}\right)\left(39{,}312\,\frac{ft^3}{hr}\right)}{\left(55.2\,\frac{ft\text{-}lbf}{lbm\text{-}°R}\right)(460 + 100°F)}$$

$$= \boxed{2173 \text{ lbm/hr}}$$

5.

$$C + O_2 \longrightarrow CO_2$$
$$12 + 32 \longrightarrow 48$$

The mass of oxygen required per lbm of carbon is

$$\frac{32}{12} = 2.67$$

$$2H_2 + O_2 \longrightarrow 2H_2O$$
$$4 + 32 \longrightarrow 36$$

$$\frac{32}{4} = 8\,\frac{lbm\,O_2}{lbm\,H_2}$$

$$S + O_2 \longrightarrow SO_2$$
$$32.1 + 32 \longrightarrow 64.1$$

$$\frac{32}{32.1} = 1\,\frac{lbm\,O_2}{lbm\,S}$$

Nitrogen does not burn.

$$(0.84)(2.67) + (0.153)(8) + (0.003)(1)$$
$$= 3.47 \text{ lbm } O_2/\text{lbm fuel}$$

Since air is 0.2315 oxygen,

$$\text{air} = \frac{3.47}{0.2315} = \boxed{15.0 \text{ lbm air/lbm fuel}}$$

6.
$$C_3H_8 + 5O_2 \longrightarrow 3CO_2 + 4H_2O$$
$$44 + 160 \longrightarrow 132 + 72$$

The excess oxygen is

$$(160)(0.2) = 32$$

Since air is 0.2315 oxygen by weight, the nitrogen is

$$\left(\frac{1 - 0.2315}{0.2315}\right)(160 + 32) = 637.4$$

The percent CO_2 by weight is

$$\frac{132}{132 + 72 + 32 + 637.4} = \boxed{0.151 \ (15.1\%) \text{ [wet]}}$$

Concentrates

1. step 1: Find the mass of oxygen in the stack gases. Assume that the stack gases are at 60°F and 14.7 psia when sampled.

$$R_{CO_2} = 35.1$$
$$R_{CO} = 55.2$$
$$R_{O_2} = 48.3$$

The partial densities are

$$\rho_{CO_2} = \frac{p}{RT} = \frac{(0.12)(14.7)(144)}{(35.1)(460 + 60)}$$
$$= 1.392 \times 10^{-2} \text{ lbm/ft}^3$$

$$\rho_{CO} = \frac{(0.01)(14.7)(144)}{(55.2)(520)} = 7.374 \times 10^{-4} \text{ lbm/ft}^3$$

$$\rho_{O_2} = \frac{(0.07)(14.7)(144)}{(48.3)(520)} = 5.9 \times 10^{-3} \text{ lbm/ft}^3$$

$$CO_2 = \frac{32}{44} = 0.7273 \text{ oxygen}$$

$$CO = \frac{16}{28} = 0.5714 \text{ oxygen}$$

O_2 is all oxygen. In 100 ft^3 of stack gases, the total oxygen weight is

$$(100)[(0.7273)(1.392 \times 10^{-2})$$
$$+ (7.374 \times 10^{-4})(0.5714) + (5.9 \times 10^{-3})(1.00)]$$
$$= 1.645 \text{ lbm } O_2/100 \text{ ft}^3$$

step 2: Since air is 23.15% oxygen by weight, the air per 100 ft^3 is

$$\frac{1.645}{0.2315} = 7.106 \text{ lbm air/100 ft}^3$$

step 3: Find the mass of carbon in the stack gases.

$$CO_2 = \frac{12}{44} = 0.2727 \text{ carbon}$$

$$CO = \frac{12}{28} = 0.4286 \text{ carbon}$$

$$(100)[(1.392 \times 10^{-2})(0.2727)$$
$$+ (7.374 \times 10^{-4})(0.4286)]$$
$$= 0.4112 \text{ lbm carbon/100 ft}^3$$

step 4: The coal is 80% carbon, so the air per lbm of coal is

$$\left(\frac{0.80 \ \dfrac{\text{lbm carbon}}{\text{lbm coal}}}{0.4112 \ \dfrac{\text{lbm carbon}}{100 \text{ ft}^3}}\right)\left(7.106 \ \frac{\text{lbm air}}{100 \text{ ft}^3}\right)$$
$$= 13.82 \text{ lbm air/lbm coal}$$

This does not include air to burn the hydrogen, since Orsat is a dry analysis.

step 5: The theoretical air for the hydrogen is

$$(34.34)\left(0.04 - \frac{0.02}{8}\right) = 1.29 \text{ lbm air/lbm coal}$$

step 6: Ignoring any excess air for the hydrogen, the air per lbm of coal is

$$13.82 + 1.29 = \boxed{15.11 \text{ lbm air/lbm coal}}$$

Alternate solution:

$$\frac{\text{lbm air}}{\text{lbm fuel}} = \frac{(3.04)(N_2)[C]}{CO_2 + CO}$$
$$= \frac{(3.04)(80\%)(0.80)}{12\% + 1\%}$$
$$= \boxed{15.0 \text{ lbm air/lbm fuel}}$$

2. step 1: The heating value per lbm of coal is

$$(0.75)(14{,}093) + \left(0.05 - \frac{0.03}{8}\right)(60{,}958)$$
$$= 13{,}389 \text{ BTU/lbm}$$

step 2: The gravimetric analysis of 1 lbm of fuel is

carbon:	0.75 lbm
free hydrogen: $\left(0.05 - \dfrac{0.03}{8}\right)$	0.0463
water: $(9)(0.05 - 0.0463)$	0.0333
nitrogen:	0.02

step 3: The theoretical stack gases per lbm of coal for 0.75 lbm of coal are

$$CO_2 = (0.75)(3.67) = 2.753$$
$$N_2 = (0.75)(8.78) = 6.585$$

All products are summarized in the following table.

	CO_2	N_2	H_2O
from C	2.753	6.583	
from H_2		1.218	0.417
from H_2O			0.0333
from O_2	shows up in CO_2, H_2O		
from N_2		0.02	
Total:	2.753	7.821	0.4503

step 4: Assume that the stack gases leave at 1000°F.

$$T_{\text{ave}} = \left(\tfrac{1}{2}\right)\left[(60+460)+(1000+460)\right]$$
$$= 990 \quad [\text{say } 1000°R]$$
$$\bar{c}_p(CO_2) = 0.251$$
$$\bar{c}_p(N_2) = 0.255$$
$$\bar{c}_p(H_2O) = 0.475$$

The energy required to raise the combustion products from 1 lbm of coal 1°F is

$$(2.753)(0.251) + (7.821)(0.255) + (0.4503)(0.475)$$
$$= 2.9 \text{ BTU}$$

step 5: Assuming all combustion heat goes into the stack gases, the final temperature is

$$T_2 = T_1 + \frac{HHV}{2.9}$$
$$= 60 + \frac{13,389}{2.9} = 4677°F \quad [\text{unreasonable}]$$

step 6: In reality, there will be approximately 40% excess air, and 75% of the heat will be absorbed by the boiler. The excess air is

$$\frac{(0.40)(7.821)}{0.7685} = 4.071 \text{ lbm}$$

$\bar{c}_p$ for air at 1000°R is approximately

$$(0.7685)(0.255) + (0.2315)(0.236) = 0.251$$

$$T_2 = 60 + \frac{(13,389)(1-0.75)}{2.9 + (4.071)(0.251)}$$

$$= \boxed{913.5°R}$$

3. *step 1:* Find the stoichiometric oxygen required per lbm of fuel oil.

C $\longrightarrow CO_2$:	$(0.8543)(2.67) =$	2.2810
H $\longrightarrow H_2O$:	$(0.1131)(8) =$	0.9048
S $\longrightarrow SO_2$:	$(0.0034)(1) =$	0.0034
less O_2 in fuel:		-0.0270
		3.1622

step 2: The theoretical nitrogen is

$$\left(\frac{3.1622}{0.2315}\right)(0.7685) = 10.497$$

The actual nitrogen with excess air and the nitrogen in the fuel is

$$0.0022 + (1.6)(10.497) = 16.8$$

This is a volume of

$$V = \frac{mRT}{p} = \frac{(16.8)(55.2)(460+60)}{(14.7)(144)}$$
$$= 227.81 \text{ ft}^3$$

step 3: The excess oxygen is

$$(0.6)(3.1622) = 1.897 \text{ lbm oxygen/lbm fuel}$$

This is a volume of

$$V = \frac{mRT}{p} = \frac{(1.897)(48.3)(460+60)}{(14.7)(144)}$$
$$= 22.51 \text{ ft}^3$$

step 4: The 60°F combustion product volumes per lbm of fuel will be

CO_2:	$(0.8543)(31.6) =$	27.0	ft^3
H_2O:	$(0.1131)(189.5) =$	21.43	ft^3
SO_2:	$(0.0034)(11.84) =$	0.04	ft^3
N_2:	from step 2 $=$	227.81	ft^3
O_2:	from step 3 $=$	22.51	ft^3
		298.79	ft^3

step 5: At 600°F, the wet volume will be

$$V_{\text{wet}} = \left(\frac{460+600°}{460+60°}\right)(298.79)$$
$$= 609.1 \text{ ft}^3$$

At 600°F, the dry volume will be

$$V_{\text{dry}} = \left(\frac{460 + 600°}{460 + 60°}\right)(298.79 - 21.43)$$

$$= 565.4 \text{ ft}^3$$

step 6: The percent carbon dioxide by volume (dry) is

$$\frac{27}{27 + 0.04 + 227.81 + 22.51} = \boxed{0.097 \ (9.7\%)}$$

4. The hydrogen and oxygen listed must both be free since the water vapor is listed separately.

step 1: The usable percent of carbon per lbm of fuel is

$$0.5145 - \frac{(2816)(0.209)}{15,395} = 0.4763$$

step 2: The theoretical oxygen required per lbm of fuel is

$$
\begin{array}{llr}
\text{C} \longrightarrow \text{CO}_2: & (0.4763)(2.67) = & 1.2717 \\
\text{H}_2 \longrightarrow \text{H}_2\text{O}: & (0.0402)(8.0) = & 0.3216 \\
\text{S} \longrightarrow \text{SO}_2: & (0.0392)(1.0) = & 0.0392 \\
\text{less O}_2 \text{ in fuel:} & & \underline{-0.0728} \\
& & 1.5597
\end{array}
$$

step 3: The theoretical air is

$$\frac{1.5597}{0.2315} = 6.737 \text{ lbm air/lbm fuel}$$

step 4: Ignoring fly ash, the theoretical dry products are

$$
\begin{array}{llr}
\text{CO}_2: & (0.4763)(3.67) = & 1.748 \\
\text{SO}_2: & (0.0392)(2.0) = & 0.0784 \\
\text{N}_2: & 0.0093 + (0.7685)(6.737) = & \underline{5.187} \\
& & 7.013
\end{array}
$$

step 5: The excess air is

$$13.3 - 7.013 = 6.287$$

step 6: The total air supplied is

$$6.287 + 6.737 = \boxed{13.02 \text{ lbm air/lbm fuel}}$$

5. *step 1:* The density of carbon dioxide at 60°F is

$$\rho = \frac{p}{RT} = \frac{(14.7)(144)}{(35.1)(460 + 60)}$$

$$= 0.1160 \text{ lbm/ft}^3$$

The carbon dioxide is $12/44 = 0.2727$ carbon, so the mass of carbon per 100 ft^3 of stack gases is

$$(0.2727)(0.1160)(100)(0.095) = 0.3005$$

step 2: The density of nitrogen is

$$\rho = \frac{(14.7)(144)}{(55.2)(460 + 60)} = 0.7375 \text{ lbm/ft}^3$$

The mass of nitrogen in 100 ft^3 is

$$(0.815)(100)(0.07375) = 6.0106 \text{ lbm}$$

step 3: The actual nitrogen per lbm of coal is

$$\left[\frac{(0.65)\left(1 - 0.03 \frac{\text{lbm carbon}}{\text{lbm fuel}}\right)}{0.3005 \frac{\text{lbm carbon}}{100 \text{ ft}^3}}\right]\left(6.0106 \frac{\text{lbm N}_2}{100 \text{ ft}^3}\right)$$

$$= 12.611 \text{ lbm N}_2/\text{lbm fuel}$$

step 4: Assuming the ash pit loss coal has the same composition as the unburned coal, the theoretical nitrogen per lbm of fuel burned is

$$(0.7685)(9.45)(1 - 0.03) = 7.0445 \text{ lbm/lbm}$$

step 5: The percent excess air is

$$\frac{12.611 - 7.0445}{7.0445} = \boxed{0.790 \ (79\%)}$$

6.
$$[\text{C}] = (1 - 0.03)(0.6734) = 0.6532$$

$$q_4 = \frac{(10,150)(0.6532)(1.6)}{15.5 + 1.6}$$

$$= \boxed{620 \text{ BTU/lbm}}$$

7. Density is volumetrically weighted.
Methane:

$$B = 0.93$$

$$\rho = \frac{p}{RT} = \frac{(14.7)(144)}{(96.4)(460 + 60)}$$

$$= 0.0422 \text{ lbm/ft}^3$$

$$\frac{\text{ft}^3 \text{ air}}{\text{ft}^3 \text{ fuel}} = 9.52$$

Products: 1 ft^3 CO$_2$, 2 ft^3 H$_2$O

$$\text{HHV} = 1013$$

Similar results for all the other fuel components are tabulated in the following table.

gas	B	ρ	ft³ air	HHV	volumes of products CO₂	H₂O	other
CH$_4$	0.93	0.0422	9.52	1013	1	2	–
N$_2$	0.034	0.0737	–	–	–	–	1N$_2$
CO	0.0045	0.0737	2.38	322	1	–	–
H$_2$	0.0182	0.0053	2.39	325	–	1	–
C$_2$H$_4$	0.0025	0.0739	14.29	1614	2	2	–
H$_2$S	0.0018	0.0900	7.15	647	–	1	1SO$_2$
O$_2$	0.0035	0.0843	–	–	–	–	–
CO$_2$	0.0022	0.1160	–	–	–	1	–

The composite density is

$$\sum B\rho = \boxed{0.0431 \text{ lbm/ft}^3}$$

The air is 20.9% oxygen by volume. The theoretical air requirements are

$$\sum B_i(\text{ft}^3 \text{ air}) - \frac{\text{oxygen in fuel}}{0.209}$$
$$= 8.9564 - \frac{0.0035}{0.209}$$
$$= \boxed{8.94 \text{ ft}^3 \text{ air/ft}^3 \text{ fuel}}$$

The theoretical oxygen will be

$$(8.94)(0.209) = 1.868 \text{ ft}^3/\text{ft}^3$$

The excess oxygen will be

$$(0.4)(1.868) = 0.747 \text{ ft}^3/\text{ft}^3$$

Similarly, the total nitrogen in the stack gases is

$$(1.4)(0.791)(8.94) + 0.034 = 9.934$$

The stack gases per ft^3 of fuel are

excess O_2:		= 0.747
N_2:		= 9.934
SO_2:		= 0.0018
CO_2:	$(0.93)(1)+(0.0045)(1)$ $+(0.0025)(2)+(0.0022)(1)$	= 0.9417
H_2O:	$(0.93)(2)+(0.0182)(1)$ $+(0.0025)(2)+(0.0018)(1)$	= 1.885

The total wet volume is 13.51 ft^3/ft^3 fuel.

The total dry volume is 11.62 ft^3/ft^3 fuel.

The volumetric analyses are

	O_2	N_2	SO_2	CO_2	H_2O
wet	0.055	0.735	–	0.070	0.140
dry	0.064	0.855	–	0.081	–

8. *step 1:* The heat absorbed in the boiler is

$$(11.12 \text{ lbm H}_2\text{0}) \left(970.3 \frac{\text{BTU}}{\text{lbm}}\right) = 10{,}789.7 \text{ BTU/lbm}$$

step 2: The losses for heating stack gases can be found as follows.

The burned carbon per lbm of fuel is

$$0.7842 - (0.315)(0.0703) = 0.7621 \text{ lbm/lbm}$$

The weight of dry fuel gas is

$$\frac{[(11)(14.0) + (8)(5.5) + (7)(0.42 + 80.08)]}{(3)(14 + 0.42)} \times \left(0.7621 + \frac{0.01}{1.833}\right)$$
$$= 13.51 \text{ lbm stack gases/lbm fuel}$$

Assume $\bar{c}_p = 0.245$ BTU/lbm-°F.

Assume that the air initially is 73°F.

$$q_1 = (13.51)(0.245)(575 - 73)$$
$$= 1661.6 \text{ BTU/lbm}$$

The coal analysis given adds to 100% without the 1.91% moisture. Assume that the H_2 and O_2 are free gases. With little error, the amount of moisture can be taken as 0.0191 lbm/lbm coal.

To heat, evaporate, and superheat the water formed from combustion, assume $\bar{c}_p = 0.46$ for superheated steam.

$$(0.0556 \text{ lbm hydrogen}) \left(9 \frac{\text{lbm water}}{\text{lbm hydrogen}}\right)$$
$$\times \left[(1)(212 - 73) + 970.3 + (0.46)(575 - 212)\right]$$
$$= 638.7 \text{ BTU/lbm}$$

To evaporate the moisture in the coal,

$$(0.0191)\left[(1)(212 - 73) + 970.3 + (0.46)(575 - 212)\right]$$
$$= 24.4 \text{ BTU/lbm}$$

From the psychrometric chart,

$$\omega = 90 \text{ grains/lbm air}$$
$$\frac{90}{7000} = 0.0129 \text{ lbm water/lbm air}$$

Since the carbon burned per lbm of fuel is

$$0.7842 - (0.315)(0.0703) = 0.7621$$

The air fuel ratio calculation is modified for sulfur content.

$$\frac{\text{lbm air}}{\text{lbm fuel}} = \frac{(3.04)(80.08)\left(0.7621 + \frac{0.01}{1.833}\right)}{14 + 0.42}$$
$$= 12.96$$

The energy to superheat the moisture in the air is approximately

$$\left(0.0129 \frac{\text{lbm water}}{\text{lbm air}}\right) \left(12.96 \frac{\text{lbm}}{\text{air}}\right) \left(0.46 \frac{\text{BTU}}{\text{lbm-°F}}\right)$$
$$\times (575°\text{F} - 73°\text{F})$$
$$= 38.6 \text{ BTU/lbm}$$

PROFESSIONAL PUBLICATIONS, INC. ● Belmont, CA

The energy lost in incomplete combustion of carbon is

$$q_4 = \frac{(10{,}143)(0.7621)(0.42)}{14 + 0.42} = 225.1 \text{ BTU/lbm}$$

The energy lost in unburned carbon is

$$q_5 = \frac{(14{,}093)(0.0703)(31.5)}{100} = 312.1$$

Energy lost in radiation and unaccounted for is

$$14{,}000 - 10{,}789.7 - 1661.6 - 638.7 - 24.4$$
$$- 38.6 - 225.1 - 312.1$$
$$= 309.8 \text{ BTU/lbm}$$

9. (a) The hardness is calculated from standard factors (see App. D.)

ion	amount as substance	factor	amount as $CaCO_3$
Ca^{++}	80.2	× 2.5 =	200.5
Mg^{++}	24.3	× 4.1 =	99.63
Fe^{++}	1.0	× 1.79 =	1.79
Al^{+++}	0.5	× 5.56 =	2.78

$$\text{hardness} = \boxed{304.7}$$

(b) To remove the carbonate hardness, first remove the carbon dioxide.

$$CO_2: \quad (19)(2.27) = 43.13 \text{ mg/}\ell \text{ as } CaCO_3$$

Add lime to remove the carbonate hardness. It does not matter if the HCO_3^- comes from Ca^{++}, Mg^{++}, Fe^{++}, or Al^{+++}. Adding lime will remove it.

There might be extra Mg^{++}, Ca^{++}, Fe^{++}, or Al^{+++} ions left over in the form of noncarbonate hardness, but the carbonate hardness is

$$HCO_3^-: \quad (185)(0.82) = 151.7 \text{ mg/}\ell \text{ as } CaCO_3$$

The total equivalents to be neutralized are

$$43.13 + 151.7 = 194.83 \text{ mg/}\ell$$

Convert these $CaCO_3$ equivalents back to $CA(OH)_2$.

$$\frac{194.83}{1.35} = \boxed{144.3 \text{ mg/}\ell \text{ of } Ca(OH)_2}$$

No soda ash is required since it is used to remove non-carbonate hardness.

10. (a) $Ca(HCO_3)_2$ and $MgSO_4$ both contribute to hardness. Since 100 mg/ℓ of hardness is the goal, leave all

$MgSO_4$ in the water. No soda ash is required. Plan to take out $137 + 72 - 100 = 109$ mg/ℓ of $Ca(HCO_3)_2$.

Including the excess, the lime required is

$$\text{pure } Ca(OH)_2 = 30 + \frac{109}{1.35} = 110.74 \text{ mg/}\ell$$

$$(110.74)(8.345) = \boxed{924 \text{ lbm/million gal}}$$

(b) The hardness removed is

$$137 + 72 - 100 = 109 \text{ mg/}\ell$$

The conversion from mg/ℓ to lbm/MG is 8.345.

$$\left(109 \frac{mg}{\ell}\right)(8.345) = 909.6 \text{ lbm hardness/million gal}$$

$$\left(\frac{0.5 \text{ lbm}}{1000 \text{ grain}}\right)\left(909.6 \frac{\text{lbm}}{\text{million gal}}\right)\left(7000 \frac{\text{grain}}{\text{lbm}}\right)$$

$$= \boxed{3184 \text{ lbm/million gal}}$$

Timed

1. (a) The silicon in ash is SiO_2 with a molecular weight of

$$28.09 + (2)(16) = 60.09$$

The oxygen tied up is

$$\left[\frac{(2)(16)}{28.09}\right](0.061) = 0.0695$$

The silicon ash produced per hour is

$$(15{,}300)(0.061 + 0.0695) = 1996.7 \text{ lbm/hr}$$

The refuse silicon is

$$(410)(1 - 0.30) = 287 \text{ lbm/hr}$$

The emission rate is

$$1996.7 - 287 = \boxed{1709.7 \text{ lbm/hr}}$$

(b) Each pound of sulfur produces 2 lbm of SO_2.

$$(15{,}300)(0.0244)(2) = \boxed{746.6 \text{ lbm/hr}}$$

(c) The heating value of the fuel is

$$(14{,}093)(0.7656) + (60{,}958)\left(0.055 - \frac{0.077}{8}\right)$$
$$+ (3983)(0.0244)$$
$$= 13{,}653 \text{ BTU/lbm}$$

The gross available combustion power is

$$\left(15{,}300 \,\frac{\text{lbm}}{\text{hr}}\right)\left(13{,}653 \,\frac{\text{BTU}}{\text{lbm}}\right) = 2.09 \times 10^8 \text{ BTUH}$$

The carbon content of the ash reduces this slightly.

$$(410)(0.30)(14{,}093) = 1.73 \times 10^6 \text{ BTUH}$$

The remaining combustion power is

$$2.09 \times 10^8 - 1.73 \times 10^6 = 2.07 \times 10^8 \text{ BTUH}$$

Losses in the steam generator and electrical generator reduce this further.

$$(0.86)(0.95)(2.07 \times 10^8) = 1.7 \times 10^8 \text{ BTUH}$$

With an electrical output of 17,000 kW, the remaining thermal energy would be removed by the cooling water.

$$\Delta T = \frac{Q}{\dot{m} c_p}$$
$$= \frac{(1.7 \times 10^8) - (17{,}000)\left(3413 \,\frac{\text{BTUH}}{\text{kW}}\right)}{(225)(62.4)(3600)(1.0)}$$
$$= \boxed{2.21^\circ\text{F}}$$

(d) The allowable emission rate is

$$\frac{\left(0.1 \,\frac{\text{lbm}}{\text{MBTU}}\right)\left(15{,}300 \,\frac{\text{lbm}}{\text{hr}}\right)\left(13{,}653 \,\frac{\text{BTU}}{\text{lbm}}\right)}{1{,}000{,}000 \,\frac{\text{BTU}}{\text{MBTU}}}$$
$$= 20.89 \text{ lbm/hr}$$

$$\eta = \frac{1709.7 - 20.89}{1709.7} = \boxed{0.988 \ (98.8\%)}$$

2. For propane,

$$C_3H_8 + 5O_2 \longrightarrow 3CO_2 + 4H_2O$$
$$\text{MW:} \quad 44.09 + 160 \longrightarrow 132.03 + 72.06$$

Oxygen is

$$(160)(1.4) = 224$$

The excess is

$$224 - 160 = 64$$

The mass ratio of nitrogen to oxygen is

$$\frac{G_N}{G_O} = \frac{B_N R_O}{B_O R_N} = \frac{(0.40)(48.3)}{(0.60)(55.2)}$$
$$= 0.583$$

The accompanying nitrogen is

$$(224)(0.583) = 130.6$$

The mass balance per mole of propane is

$$C_3H_8 + O_2 + N_2 \longrightarrow CO_2 + H_2O + O_2 + N_2$$
$$44.09 + 224 + 130.6 \longrightarrow 132.03 + 72.06 + 64 + 130.6$$

At standard industrial conditions (60°F, 1 atm), the propane density is

$$\rho = \frac{p}{RT} = \frac{(14.7)(144)}{(35.0)(460 + 60)}$$
$$= 0.1163 \text{ lbm/ft}^3$$

Since the flow rate is 250 SCFM, the mass flow rate of the propane is

$$(250)(0.1163) = 29.075 \text{ lbm/min}$$

Scaling the other mass balance factors down,

$$C_3H_8 + O_2 + N_2 \longrightarrow CO_2 + H_2O + O_2 + N_2$$
$$29.08 + 147.72 + 86.1 \longrightarrow 87.07 + 47.52 + 42.20 + 86.1$$

$$\boxed{\text{The oxygen flow is 147.72 lbm/min.}}$$

The specific volumes of the reactants are

$$v_{C_3H_8} = \frac{RT}{p} = \frac{(35.0)(460 + 80)}{(14.7)(144)}$$
$$= 8.929 \text{ ft}^3/\text{lbm}$$

$$v_{O_2} = \frac{(48.3)(540)}{(14.7)(144)} = 12.321 \text{ ft}^3/\text{lbm}$$

$$v_{N_2} = \frac{(55.2)(540)}{(14.7)(144)} = 14.082 \text{ ft}^3/\text{lbm}$$

The total incoming volume is

$$(29.08)(8.929) + (147.72)(12.321) + (86.1)(14.082)$$
$$= 3292 \text{ ft}^3/\text{min}$$

Since velocity must be kept below 400 ft/min,

$$A = \frac{Q}{v} = \frac{3292}{400} = \boxed{8.23 \text{ ft}^2}$$

Similarly, the specific volumes of the products are

$$v_{CO_2} = \frac{(35.1)(460+460)}{(8)(144)} = 28.03 \text{ ft}^3/\text{lbm}$$

$$v_{H_2O} = \frac{(85.8)(920)}{1152} = 68.52 \text{ ft}^3/\text{lbm}$$

$$v_{O_2} = \frac{(48.3)(920)}{1152} = 38.57 \text{ ft}^3/\text{lbm}$$

$$v_{N_2} = \frac{(55.2)(920)}{1152} = 44.08 \text{ ft}^3/\text{lbm}$$

The total exhaust volume is

$$(87.07)(28.03) + (47.52)(68.52) + (42.20)(38.57)$$
$$+ (86.1)(44.08)$$
$$= \boxed{11,120 \text{ ft}^3/\text{min}}$$

$$A_{stack} = \frac{11,120}{800} = \boxed{13.9 \text{ ft}^2}$$

For ideal gases, the partial pressure is volumetrically weighted. The water vapor partial pressure is

$$(8)\left[\frac{(47.52)(68.52)}{11,120}\right] = 2.34 \text{ psia}$$

The saturation temperature corresponding to 2.34 psia is

$$\boxed{T_{\text{dew point}} = 132°F}$$

3. (a) Assume that the nitrogen and oxygen can be varied independently.

$$C_3H_8 + 5O_2 \longrightarrow 3CO_2 + 4H_2O$$
weights: 44.09 + 160 ⟶ 132.03 + 72.06
moles: (1)   (5)   (3)   (4)

Assume 0% excess oxygen. (This is justified since enthalpy increase information is not given for oxygen.)

Use the enthalpy of formation data to calculate the heat of reaction (maximum flame temperature). Since oxygen and nitrogen have $\Delta H_f = 0$,

$$(3)(-169,300) + (4)(-104,040) - 28,800$$
$$= -952,860 \text{ BTU/mole}$$

The negative sign indicates an exothermal reaction. Let x be the number of moles of nitrogen per mole of propane. Use the nitrogen to cool the combustion.

$$952,860 = (3)(39,791) + (4)(31,658) + x(24,471)$$
$$x = 28.89 \text{ moles}$$

$$M_{N_2} = (28.89)(28.016) = \boxed{809.4 \text{ lbm/mole}}$$

$$M_{O_2} = \boxed{160 \text{ lbm/mole}}$$

Each pmole of propane produces 4 moles of water.

$$m_{H_2O} = (4)(18) = 72 \text{ lbm/mole}$$

(All flows are per pmole propane.)

(b) The partial pressure is volumetrically weighted. This is the same molar weighting.

product	moles	volumetric fraction
CO₂	3.0	0.0836
H₂O	4.0	0.1115
N₂	28.89	0.8049
O₂	0	0
	35.89	1.000

The partial pressure of the water vapor is

$$p_{H_2O} = (0.1115)(14.7) = 1.64 \text{ psia}$$

This corresponds to approximately 118°F.

Since $T_{stack} = 100°F$, the maximum vapor pressure is 0.9492 psia.

Let n be the number of moles of water in the stack gas.

$$\frac{0.9492}{14.7} = \frac{n}{35.89}$$
$$n = 2.317 \text{ moles}$$

The liquid water removed is

$$4 - 2.317 = \boxed{1.683 \text{ moles of H}_2\text{O/mole C}_3\text{H}_8}$$

Expressing this result in lbm,

$$(1.683)\left(18 \frac{\text{lbm}}{\text{mole}}\right) = \boxed{30.29 \text{ lbm H}_2\text{O/mole C}_3\text{H}_8}$$

HEAT TRANSFER

Warmups

1. Use the value of k at the average temperature of $\left(\frac{1}{2}\right)(150 + 350) = 250°F$.

$$k = (0.030)[1 + (0.0015)(250)] = \boxed{0.04125}$$

2.
$$q = \frac{kA\Delta T}{L}$$
$$= \frac{\left(0.038\,\frac{\text{BTU-ft}}{\text{hr-ft}^2\text{-°R}}\right)(350°F)}{1\,\text{ft}}$$
$$= \boxed{13.3\ \text{BTU/hr-ft}^2}$$

3. $\Delta T_A = 190°F - 85°F = 105°F$

$\Delta T_B = 160°F - 85°F = 75°F$

$$\Delta T_m = \frac{105 - 75}{\ln\left(\dfrac{105}{75}\right)} = \boxed{89.16°F}$$

4. (a) h should be evaluated at $\left(\frac{1}{2}\right)(T_{\text{pipe}} + T_{\text{fluid}})$. The midpoint pipe temperature is

$$\left(\tfrac{1}{2}\right)(160°F + 190°F) = 175°F$$

At the start of the heating process,

$$T_f = \left(\tfrac{1}{2}\right)(85°F + 175°F) = \boxed{130°F}$$

(b) The initial film coefficient should be evaluated at

$$T = \left(\tfrac{1}{2}\right)(190°F + 160°F) = \boxed{175°F}$$

5. Assume counter-flow operation to find ΔT_M.

$$
\begin{array}{c}
55° \xrightarrow{\hspace{1cm}\text{oil}\hspace{1cm}} 87° \\
\text{A} \hspace{3cm} \text{B} \\
270° \xleftarrow{\hspace{1cm}\text{gases}\hspace{1cm}} 350°
\end{array}
$$

$\Delta T_A = 270°F - 55°F = 215°F$

$\Delta T_B = 350°F - 87°F = 263°F$

$$\Delta T_m = \frac{215°F - 263°F}{\ln\left(\dfrac{215}{263}\right)} = \boxed{238.19°F}$$

6. If the steam is 87% wet, the quality is $x = 0.13$.

$$v = 0.01727 + (0.13)(8.515 - 0.01727)$$
$$= 1.122\ \text{ft}^3/\text{lbm}$$
$$\rho = \frac{1}{v} = \frac{1}{1.122} = \boxed{0.891\ \text{lbm/ft}^3}$$

7.
$$\mu = \left(0.458 \times 10^{-3}\,\frac{\text{lbm}}{\text{ft-sec}}\right)\left(3600\,\frac{\text{sec}}{\text{hr}}\right)$$
$$= \boxed{1.6488\ \text{lbm/ft-hr}}$$

8. Assuming that the walls are reradiating, non-conducting, and vary in temperature from 2200°F at the inside to 70°F at the outside, Fig. 10.15, curve 6 can be used to find $F_{1\text{-}2}$ using $x = 3\,\text{in}/6\,\text{in} = 0.5$, $F_{1\text{-}2} = 0.38$.

$$q = AE$$
$$= \left(\frac{3\,\text{in}}{12\,\dfrac{\text{in}}{\text{ft}}}\right)^2 \left(0.1713 \times 10^{-8}\,\frac{\text{BTU}}{\text{hr-ft}^2\text{-°R}}\right)(0.38)$$
$$\times\ [(2200 + 460)^4 - (70 + 460)^4]$$
$$= \boxed{2033.6\ \text{BTU/hr}}$$

9.
$$U = \frac{1}{\sum\left(\dfrac{L_i}{k_i}\right) + \sum\dfrac{1}{h_j}}$$
$$L = \frac{4\,\text{in}}{12\,\dfrac{\text{in}}{\text{ft}}} = 0.333\ \text{ft}$$
$$k = 13.9\ \text{BTU-ft/ft}^2\text{-hr-°F}$$
$$h_{\text{inside}} \approx 1.65$$

h_{outside} is more difficult. Assuming linearity between 0 and 15 mph,

$$h_{\text{outside}} \approx 1.65 + \left(\frac{10}{15}\right)(6.00 - 1.65) = 4.55$$
$$U = \frac{1}{\left(\dfrac{0.333}{13.9}\right) + \dfrac{1}{1.65} + \dfrac{1}{4.55}}$$
$$= \boxed{1.18\ \text{BTU/ft}^2\text{-hr-°F}}$$

10.

$$\text{Re} = \frac{Dv}{\nu}$$

$$D = \frac{0.6 \text{ in}}{12 \frac{\text{in}}{\text{ft}}} = 0.05 \text{ ft}$$

$$v = 2 \text{ ft/sec}$$

Assuming SAE 10W at 100°F, $\nu = 45$ centistokes (p. 4-27, 15th edition, *Cameron Hydraulic Data*).

$$(45 \text{ centistokes}) \left(\frac{1 \text{ stoke}}{100 \text{ centistoke}} \right) \left(\frac{1 \text{ ft}^2}{929 \text{ sec-stoke}} \right)$$

$$= 4.84 \times 10^{-4} \text{ ft}^2/\text{sec}$$

$$\text{Re} = \frac{(0.05 \text{ ft}) \left(2 \frac{\text{ft}}{\text{sec}} \right)}{4.84 \times 10^{-4} \frac{\text{ft}^2}{\text{sec}}} = \boxed{206.6}$$

Concentrates

1. Since the wall temperatures are given (i.e., temperatures are not environment temperatures) it is not necessary to consider films. On a 1 ft^2 basis,

$$q = \frac{(1 \text{ ft}^2)(1000°\text{F} - 200°\text{F})}{\dfrac{\frac{3}{12}}{0.06} + \dfrac{\frac{5}{12}}{0.5} + \dfrac{\frac{6}{12}}{0.8}}$$

$$= 142.2 \text{ BTU/hr}$$

Solving for ΔT up to the first interface,

$$\Delta T = \frac{\left(142.2 \frac{\text{BTU}}{\text{hr}} \right) \left(\dfrac{\frac{3}{12} \text{ ft}}{0.06 \frac{\text{BTU-ft}}{\text{ft}^2\text{-hr-}°\text{F}}} \right)}{1 \text{ ft}^2}$$

$$= 592.5°\text{F}$$

The temperature at the first interface is

$$T = 1000°\text{F} - 592.5°\text{F} = 407.5°\text{F}$$

Similarly for the other interface,

$$\Delta T = \frac{(142.2) \left(\dfrac{\frac{6}{12}}{0.8} \right)}{1} = 88.9°\text{F}$$

$$T = 200°\text{F} + 88.9°\text{F} = \boxed{288.9°\text{F}}$$

2.

$$r_a = \frac{d}{2} = \frac{3.5 \text{ in}}{(2) \left(12 \frac{\text{in}}{\text{ft}} \right)} = 0.1458 \text{ ft}$$

$$r_b = \frac{4.0}{(2)(12)} = 0.1667 \text{ ft}$$

$$r_c = r_b + t_{\text{insulation}}$$

$$= 0.1667 + \frac{2}{12} = 0.3334 \text{ ft}$$

Interpolating for steel pipe, $k \approx 25.4$. Initially assume $h_o = 1.5$ BTU/hr-ft^2-°F.

Either the Seider-Tate or the Graetz correlation can be used to find h_i. For fully-developed laminar flow (i.e., a long distance down a long pipe), the Nusselt number is

$$\text{Nu} = \frac{h_i D}{k} \approx 3.66$$

At 350°F, $k_{\text{air}} \approx 0.0203$ BTU/hr-ft^2-°F.

$$h_i = \frac{k \text{Nu}}{D} = \frac{(0.0203)(3.66)}{(2)(0.1458)} = 0.255$$

Neglect thermal resistance between pipe and insulation.

$$q = \frac{2\pi(100 \text{ ft})(350°\text{F} - 50°\text{F})}{\dfrac{1}{(0.1458)(0.255)} + \dfrac{\ln \left(\frac{0.1667}{0.1458} \right)}{25.4}} $$

$$+ \dfrac{\ln \left(\frac{0.3334}{0.1667} \right)}{0.05} + \dfrac{1}{(0.3334)(1.5)}$$

$$= \frac{188,496}{26.90 + 0.00527 + 13.863 + 2.00}$$

$$= \boxed{4407 \text{ BTU/hr}}$$

Greater accuracy requires finding h_o. Solve for ΔT up to the outer film to find the insulation surface temperature.

$$\Delta T = \left[\frac{4407 \frac{\text{BTU}}{\text{hr}}}{(2\pi \text{ ft})(100 \text{ ft})} \right]$$

$$\times \left[26.90 \frac{\text{hr-ft}^2\text{-}°\text{F}}{\text{BTU}} + 0.00527 \frac{\text{hr-ft}^2\text{-}°\text{F}}{\text{BTU}} \right.$$

$$\left. + 13.863 \frac{\text{hr-ft}^2\text{-}°\text{F}}{\text{BTU}} \right]$$

$$= 286°\text{F}$$

$$T_s \approx 350°\text{F} - 286°\text{F} = 64°\text{F}$$

The outer film should be evaluated at

$$T_f = \left(\tfrac{1}{2}\right)(T_s + T_\infty)$$
$$= \left(\tfrac{1}{2}\right)(64°F + 50°F)$$
$$= 57°F$$

For 57°F air,

$$Pr = 0.72$$
$$\frac{g\beta\rho^2}{\mu^2} \approx 2.64 \times 10^6$$

Using $L = 2r_c$ is the outside diameter in feet,

$$Gr = L^3 \Delta T \left(\frac{g\beta\rho^2}{\mu^2}\right)$$
$$= (0.6667)^3(64 - 50)(2.64 \times 10^6)$$
$$= 1.1 \times 10^7$$
$$PrGr = (0.72)(1.1 \times 10^7)$$
$$= 7.9 \times 10^6$$

$$h_o \approx (0.27)\left(\frac{\Delta T}{L}\right)^{0.25}$$
$$= (0.27)\left(\frac{64 - 50}{0.6667}\right)^{0.25} = 0.578$$

At the second iteration, the heat flow is

$$q = \frac{188{,}496}{26.90 + 0.00527 + 13.863 + \dfrac{1}{(0.3334)(0.578)}}$$

$$= \boxed{4102 \text{ BTU/hr}}$$

Additional iterations will improve the accuracy further.

3. Since no information was given about the pipe type, material, or thickness, it can be assumed that the pipe resistance is negligible.

$$T_{\text{pipe}} = T_{\text{sat}}$$

For 300 psia steam, $T_{\text{sat}} = 417.33°F$. When a vapor condenses, the vapor and condensed liquid are at the same temperature. Therefore, the entire pipe is assumed to be at 417.33°F.

The outside film should be evaluated at $\left(\tfrac{1}{2}\right)(417.33 + 70) = 243.7°F$

$$Pr = 0.715$$
$$\frac{g\beta\rho^2}{\mu^2} = 0.673 \times 10^6$$
$$Gr = \frac{L^3 g\beta\rho^2 \Delta T}{\mu^2}$$
$$= \left(\frac{4}{12}\right)^3(0.673 \times 10^6)(417.33 - 70)$$
$$= 8.66 \times 10^6$$

$$PrGr = (0.715)(8.66 \times 10^6)$$
$$= 6.2 \times 10^6$$

$$h_o \approx (0.27)\left(\frac{\Delta T}{L}\right)^{0.25} = (0.27)\left(\frac{417.33 - 70}{\frac{4}{12}}\right)^{0.25}$$
$$= 1.53$$

$$q = \frac{2\pi(50)(417.33 - 70)}{\dfrac{1}{\left(\frac{2}{12}\right)(1.53)}}$$
$$= 2.782 \times 10^4 \text{ BTU/hr}$$

To determine the radiation loss, assume oxidized steel pipe, completely enclosed.

$$F_a = 1$$
$$T_\infty = 70°F = 530°R$$
$$\epsilon_{\text{steel}} = 0.80$$

With $F_e = \epsilon_{\text{steel}}$ and $F_a = 1$,

$$q = EA$$
$$= (0.1713 \times 10^{-8})(0.80)(1)$$
$$\times \left[(\pi)\left(\frac{4}{12}\right)(50)\right]$$
$$\times \left[(417.33 + 460)^4 - (530)^4\right]$$
$$= 36{,}850 \text{ BTU/hr}$$

The total heat loss is

$$27{,}820 + 36{,}850 = 64{,}670 \text{ BTU/hr}$$

The enthalpy decrease per pound is

$$\frac{64{,}670 \ \dfrac{\text{BTU}}{\text{lbm}}}{5000 \ \dfrac{\text{lbm}}{\text{hr}}} = 12.93 \text{ BTU/lbm}$$

This is a quality loss of

$$\Delta x = \frac{\Delta h}{h_{fg}} = \frac{12.93}{809}$$

$$= \boxed{0.016 \ (1.6\%)}$$

4. The heat loss is

$$\left(8 \; \frac{W}{ft}\right)\left(3.413 \; \frac{BTU}{hr\text{-}W}\right) = 27.3 \; BTU/hr\text{-}ft$$

For 100°F film, Pr = 0.72.

$$\frac{g\beta\rho^2}{\mu^2} = 1.76 \times 10^6$$

$$Gr = L^3 \Delta T \left(\frac{g\beta\rho^2}{\mu^2}\right)$$

$$\Delta T = T_{\text{wire}} - 60°F$$

T_{wire} is unknown, so assume $T_{\text{wire}} = 150°F$.

$$Gr = \left(\frac{0.6}{12}\right)^3 (150 - 60)(1.76 \times 10^6)$$

$$= 1.98 \times 10^4$$

$$PrGr = (0.72)(1.98 \times 10^4)$$

$$= 1.43 \times 10^4$$

$$h_o \approx (0.27)\left(\frac{\Delta T}{L}\right)^{0.25} = (0.27)\left(\frac{150 - 60}{\frac{0.6}{12}}\right)^{0.25}$$

$$= 1.76$$

$$T_{\text{wire}} = \frac{q}{hA} + T_\infty = \frac{27.3}{(1.76)\left(\frac{0.6}{12}\right)\pi(1)} + 60$$

$$= 158.7°F$$

Try $T_{\text{wire}} = 158°F$.

$$Gr = \left(\frac{0.6}{12}\right)^3 (158 - 60)(1.76 \times 10^6) = 21,560$$

$$PrGr = (0.72)(21,560) = 15,523$$

$$h_o = (0.27)\left(\frac{\Delta T}{L}\right)^{0.25} = (0.27)\left(\frac{158 - 60}{\frac{0.6}{12}}\right)^{0.25}$$

$$= 1.80$$

$$T_{\text{wire}} = \frac{27.3}{(1.80)\left(\frac{0.6}{12}\right)\pi(1)} + 60$$

$$= \boxed{156.6°F \quad [\text{say } 157°F]}$$

5. The bulk temperature of the water is

$$\left(\tfrac{1}{2}\right)(70° + 190°) = 130°$$

For 130° water,

$$c_p = 0.999$$

$$\nu = 0.582 \times 10^{-5}$$

$$Pr = 3.45$$

$$k = 0.376$$

The required heat is

$$q = mc_p\Delta T = (2940)(0.999)(190 - 70)$$

$$= 352,447 \; BTU/hr$$

$$Re = \frac{vD}{\nu} = \frac{\left(3 \; \frac{ft}{sec}\right)\left(\frac{0.9}{12} \; ft\right)}{0.582 \times 10^{-5} \; \frac{ft^2}{sec}}$$

$$= 3.87 \times 10^4$$

$$h_i = \frac{(0.376)(0.0225)(3.87 \times 10^4)^{0.8}(3.45)^{0.4}}{\frac{0.9}{12}}$$

$$= 866$$

Next, find h_o. 134 psia steam has $T_{\text{sat}} = 350°F$. Assume $T_{sv} - T_s = 20°F$. Therefore, the wall is at $350 - 20 = 330°F$, and the film should be evaluated at $\left(\tfrac{1}{2}\right)(330 + 350) = 340°F$.

For 340°F steam,

$$\rho_f = \frac{1}{v_f} = \frac{1}{0.01787}$$

$$= 55.96 \; lbm/ft^3$$

$$\rho_g = \frac{1}{3.788} = 0.26 \; lbm/ft^3$$

For 350°F steam,

$$h_{fg} = 870.7 \; BTU/lbm$$

For 340°F water, $k \approx 0.392$.

$$\mu_f = \left(0.109 \times 10^{-3} \; \frac{lbm}{ft\text{-}sec}\right)\left(3600 \; \frac{sec}{hr}\right)$$

$$= 0.392 \; lbm/ft\text{-}hr$$

Using 4.17×10^8 to convert ft/sec^2 to ft/hr^2,

$$h_o = (0.725)$$

$$\times \left[\frac{(55.96)(55.96 - 0.26)(4.17 \times 10^8)(870.7)(0.392)^3}{\left(\frac{1}{12}\right)(0.392)(350 - 330)}\right]^{0.25}$$

$$= 2317 \; BTU/hr\text{-}ft^2\text{-}°F$$

$$r_o = \frac{1}{(2)(12)} = 0.0417 \; ft$$

$$r_i = \frac{0.9}{(2)(12)} = 0.0375 \; ft$$

Assuming pure copper pipes, $k_{\text{copper}} = 216$.

$$U_o = \frac{1}{\frac{1}{2317} + \left(\frac{0.0417}{216}\right)\ln\left(\frac{0.0417}{0.0375}\right) + \frac{0.0417}{(0.0375)(866)}}$$

$$= 576.0 \; BTU/hr\text{-}ft^2\text{-}°F$$

For cross-flow operation,

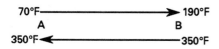

$$\Delta T_A = 350 - 70 = 280°F$$

$$\Delta T_B = 350 - 190 = 160°F$$

$$\Delta T_m = \frac{280 - 160}{\ln\left(\frac{280}{160}\right)} = 214.4°F$$

$$A_o = \frac{q}{U_o \Delta T_m} = \frac{352{,}447}{(576.0)(214.4)}$$

$$= \boxed{2.85 \text{ ft}^2}$$

6. This is a transient problem.

$$C_e = 100{,}000 \text{ BTU}/°F$$

$$R_e = \frac{1}{6500} = 0.0001538 \text{ BTU/hr-}°F$$

$$T_{8\,\text{hr}} = 40 + (70 - 40)\exp\left[\frac{-8}{(100{,}000)(0.0001538)}\right]$$

$$= \boxed{57.8°F}$$

7. Since the duct outside temperature is known, k_{duct} and h_i are not needed. Evaluate h_o at $\left(\frac{1}{2}\right)(80°F + 200°F) = 140°F$.

$$\text{Pr} = 0.72$$

$$\frac{g\beta\rho^2}{\mu^2} = 1.396 \times 10^6$$

$$\text{Gr} = L^3 \Delta T \left(\frac{g\beta\rho^2}{\mu^2}\right)$$

$$= \left(\frac{9}{12}\right)^3 (200 - 80)(1.396 \times 10^6)$$

$$= 7.07 \times 10^7$$

$$\text{PrGr} = (0.72)(7.07 \times 10^7) = 5.09 \times 10^7$$

$$h = (0.27)\left(\frac{\Delta T}{L}\right)^{0.25} = (0.27)\left(\frac{200 - 80}{\frac{9}{12}}\right)^{0.25}$$

$$= 0.96$$

The duct area per foot of length is

$$A = pL = \left(\frac{9}{12}\right)\pi(1) = 2.356 \text{ ft}^2$$

The convection losses are

$$q_{\text{convection}} = hA\Delta T = (0.96)(2.356)(200 - 80)$$

$$= 271.4 \text{ BTU/hr-ft}$$

Assume $\epsilon_{\text{duct}} \approx 0.97$. Then $F_e = \epsilon_{\text{duct}} = 0.97$. $F_A = 1$ since the duct is enclosed. The radiation losses are

$$q_{\text{radiation}} = (2.356)(0.1713 \times 10^{-8})(0.97)$$

$$\times \left[(200 + 460)^4 - (70 + 460)^4\right]$$

$$= 433.9 \text{ BTU/hr-ft}$$

$$q_{\text{total}} = 271.4 + 433.9$$

$$= \boxed{705.3 \text{ BTU/hr-ft length}}$$

8. 1 ft of rod has a volume of

$$V = \left(\frac{0.4}{12}\right)^2 \left(\frac{\pi}{4}\right)(1) = 8.727 \times 10^{-4} \text{ ft}^3/\text{ft}$$

The heat output per foot of rod is

$$q = \left(8.727 \times 10^{-4} \frac{\text{ft}^3}{\text{ft}}\right)\left(4 \times 10^7 \frac{\text{BTU}}{\text{hr-ft}^3}\right)$$

$$= 3.491 \times 10^4 \text{ BTU/hr-ft}$$

The surface temperature of the cladding is

$$T_s = \frac{q}{hA} + T_\infty$$

$$A = \pi\left(\frac{0.44}{12}\right)(1) = 0.1152$$

$$T_s = \frac{3.491 \times 10^4}{(10{,}000)(0.1152)} + 500 = 530.3$$

For the cladding,

$$r_o = \frac{0.2 + 0.02}{12} = 0.01833 \text{ ft}$$

$$r_i = \frac{0.2}{12} = 0.01667$$

$$k \approx 10.9$$

$$\Delta T = \frac{q\ln\left(\frac{r_o}{r_i}\right)}{2\pi kL}$$

$$= \frac{(3.49 \times 10^4)\ln\left(\frac{0.01833}{0.01667}\right)}{2\pi(10.9)(1)}$$

$$= 48.4°F$$

$$T_{\text{inside cladding}} = T_{\text{outside fuel rod}}$$

$$= 530.3 + 48.4 = 578.7°F$$

$$T_{\text{center}} = T_o + \frac{r_o^2 q^*}{4k}$$

$$= 578.7 + \frac{\left(\frac{0.2}{12}\right)^2 (4 \times 10^7)}{(4)(1.1)}$$

$$= \boxed{3104^\circ \text{F}}$$

9. Consider this an infinite cylindrical fin with

$$T_s = 450^\circ \text{F}$$
$$T_\infty = 80^\circ \text{F}$$
$$h = 3$$
$$p = \frac{\pi\left(\frac{1}{16}\right)}{12} = 0.01636 \text{ ft}$$
$$k \approx 215$$
$$A = \left(\frac{\pi}{4}\right)\left[\frac{1}{(16)(12)}\right]^2 = 2.131 \times 10^{-5}$$

With 2 fins joined at the middle,

$$q = 2\sqrt{(3)(0.01636)(215)(2.131 \times 10^{-5})}$$
$$\times (450 - 80)$$

$$= \boxed{11.1 \text{ BTU/hr} \quad [\text{This ignores radiation.}]}$$

10. This is a single tube in cross flow.

$$\frac{hD}{k} = C(\text{Re})^n$$

The film is evaluated at

$$\left(\tfrac{1}{2}\right)(100 + 150) = 125$$

For air at 125°F,

$$k = 0.0159$$
$$\nu = 0.195 \times 10^{-3}$$
$$\text{Re} = \frac{\text{v}D}{\nu} = \frac{(100)\left(\frac{0.35}{12}\right)}{0.195 \times 10^{-3}} = 14{,}957$$

Based on Re, C and n can be found. $C = 0.174$ and $n = 0.618$.

$$h = \frac{(0.0159)(0.174)(14{,}957)^{0.618}}{\frac{0.35}{12}}$$

$$= \boxed{36.07 \text{ BTU/hr-ft}^2\text{-}^\circ\text{F}}$$

Timed

1. Assume that the exposed wall temperatures are equal to the respective ambient temperatures.

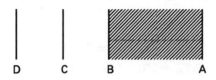

$$T_A = 80^\circ\text{F} = 540^\circ\text{R}$$
$$\frac{q_{\text{A-B}}}{A} = \frac{k\Delta T}{L} = \frac{(0.025)(540 - T_B)}{\frac{4 \text{ in}}{12 \frac{\text{in}}{\text{ft}}}}$$

$$= 40.5 - 0.75\,T_B \qquad [\text{Eq. 1}]$$

Since the spaces are evacuated, only radiation should be considered for B–C and C–D.

$$E_{\text{B-C}} = \frac{q_{\text{B-C}}}{A} = \sigma F_e F_a \left[T_B^4 - T_C^4\right]$$

Since the freezer is assumed large, $F_A = 1$.

$$F_e = \frac{(0.5)(0.1)}{0.1 + (1 - 0.1)(0.5)}$$
$$= 0.0909$$

$$E_{\text{B-C}} = (0.1713 \times 10^{-8})(0.0909)(1)$$
$$\times \left[T_B^4 - T_C^4\right]$$
$$= (1.56 \times 10^{-10})T_B^4 - (1.56 \times 10^{-10})T_C^4$$
$$[\text{Eq. 2}]$$

$$E_{\text{C-D}} = (0.1713 \times 10^{-8})(0.0909)(1)\left[T_C^4 - (460 - 60)^4\right]$$
$$= (1.56 \times 10^{-10})T_C^4 - 3.99 \qquad [\text{Eq. 3}]$$

Eq. 2 = Eq. 3.

$$(1.56 \times 10^{-10})T_B^4$$
$$- (1.56 \times 10^{-10})T_C^4 = (1.56 \times 10^{-10})T_C^4 - 3.99$$
$$T_B^4 - T_C^4 = T_C^4 - 2.56 \times 10^{10}$$
$$T_B^4 = 2T_C^4 - 2.56 \times 10^{10}$$
$$T_C^4 = \tfrac{1}{2}T_B^4 + 1.28 \times 10^{10}$$
$$[\text{Eq. 4}]$$

Eq. 1 = Eq. 2.

$$40.5 - 0.075T_B = (1.56 \times 10^{-10})T_B^4 - (1.56 \times 10^{-10})T_C^4$$
$$T_C^4 = T_B^4 + (4.81 \times 10^8)T_B - 2.6 \times 10^{11}$$
$$[\text{Eq. 5}]$$

Eq. 4 = Eq. 5.

$$\tfrac{1}{2}T_B^4 + \left(1.28 \times 10^{10}\right) = T_B^4 + (4.81 \times 10^8)T_B$$
$$- 2.6 \times 10^{11}$$

$$T_B^4 + (9.62 \times 10^8)T_B = 5.456 \times 10^{11}$$

By trial and error, $T_B = 501.4°$R.

From Eq. 1,

$$\frac{q}{A} = 40.5 - (0.075)(501.4) = 2.895 \text{ BTU/hr-ft}^2$$

From Eq. 4,

$$T_C = \sqrt[4]{\left(\tfrac{1}{2}\right)(501.4)^4 + 1.28 \times 10^{10}} = 459°\text{R}$$

Check the intital assumption about wall temperature. Assume a film of 2 BTU/ft²-hr-°F exists.

$$T_s - T_\infty = \frac{q}{Ah} = \frac{2.9}{(1)(2)} = \boxed{1.5°\text{F}}$$

Since this is small, the assumption is valid.

2. (a) The exposed duct area is

$$\left[\frac{18 \text{ in} + 18 \text{ in} + 12 \text{ in} + 12 \text{ in}}{12 \dfrac{\text{in}}{\text{ft}}}\right](50 \text{ ft}) = 250 \text{ ft}^2$$

The duct is non-circular. The inside film coefficient cannot be evaluated without knowing the equivalent diameter of the duct.

$$D_e = \frac{(2)(18)(12)}{(18 + 12)\left(12 \dfrac{\text{in}}{\text{ft}}\right)} = 1.2 \text{ ft}$$

Check the Reynolds number. For 100°F air,

$$\nu = 0.180 \times 10^{-3}$$

$$\text{Re} = \frac{\text{v}D}{\nu} = \frac{\left(\dfrac{800 \dfrac{\text{ft}}{\text{min}}}{60 \dfrac{\text{sec}}{\text{min}}}\right)(1.2 \text{ ft})}{0.180 \times 10^{-3} \dfrac{\text{ft}^2}{\text{sec}}}$$

$$= 8.9 \times 10^4$$

This is turbulent. At 100°F,

$$\rho = 0.071 \text{ lbm/ft}^3$$

$$G = \text{v}\rho = \frac{(800)(0.071)}{60}$$

$$= 0.947 \text{ lbm/sec-ft}^2$$

$$C = 0.00351 + (0.000001583)(100)$$

$$= 0.003668$$

$$h_i = \frac{(0.003668)[(3600)(0.947)]^{0.8}}{(1.2)^{0.2}}$$

$$= 2.37$$

Disregarding the duct thermal resistance, the total heat transfer coefficient is

$$\frac{1}{U} = \frac{1}{h_i} + \frac{1}{h_o} = \frac{1}{2.37} + \frac{1}{2} = 0.922$$

$$U = \frac{1}{0.922} = 1.08$$

$$q = UA(T_{\text{ave}} - T_\infty) = UA\left[\left(\tfrac{1}{2}\right)(T_{\text{in}} + T_{\text{out}}) - T_\infty\right]$$

Since T_{out} is unknown, assume $T_{\text{out}} \approx 95°$F.

$$q = \left(1.08 \frac{\text{BTU}}{\text{hr-ft}^2\text{-}°\text{F}}\right)(250 \text{ ft})\left(\frac{100 + 95°\text{F}}{2} - 70°\text{F}\right)$$

$$= \boxed{7425 \text{ BTU/hr}}$$

Notice that ΔT (not ΔT_M) is used. This is idiosyncratic to the HVAC industry.

(b) The flow per hour is

$$3600GA = (3600)(0.947)\left[\frac{(12)(18)}{144}\right]$$

$$= 5113.8 \text{ lbm/hr}$$

$$c_p = 0.240$$

$$\Delta T = \frac{q}{mc_p} = \frac{7425}{(5113.8)(0.240)}$$

$$= 6.0°\text{F}$$

$$T_{\text{out}} = 100 - 6.0$$

$$= \boxed{94°\text{F} \quad [\text{close enough}]}$$

(c) Assume clean galvanized ductwork with $\epsilon = 0.0005$ and about 40 joints per 100 ft. The equivalent diameter is

$$D_e = (1.3)\left(\frac{[(12)(18)]^{0.625}}{[12 + 18]^{0.25}}\right)$$

$$= 16 \text{ in}$$

$$\Delta p = 0.057 \text{ in water/100 ft}$$

This duct system has a loss of

$$\left(\frac{50}{100}\right)(0.057) = \boxed{0.029 \text{ in w.g.}}$$

3. This is a transient problem. Check the Biot modulus to see if the lumped parameter method can be used.

$$L = \frac{V}{A_s} = \frac{\frac{4}{3}\pi r^3}{4\pi r^2} = \frac{r}{3}$$

For the largest ball,

$$L = \frac{1.5}{(3)(2)(12)} = 0.0208 \text{ ft}$$

Evaluate k for steel at

$$\left(\tfrac{1}{2}\right)(1800 + 250) \approx 1000°\text{F}$$

For mild steel, $k \approx 22$. (k for a high alloy steel is about 12–15.)

$$\text{Bi} = \frac{hL}{k} = \frac{(56)(0.0208)}{22} = 0.053$$

For smaller balls, Bi will be even smaller.

Since Bi < 0.10, the lumped parameter method can be used. The assumptions are

- homogeneous body temperature
- minimal radiation losses
- oil bath remains at 110°F
- constant h (no vaporization of oil)

$$R_e C_e = \frac{c_p \rho V}{h A_s} = \frac{c_p \rho r}{3h}$$

Use $\rho = 490$ lbm/ft^3 and $c_p = 0.11$ even those are for 32°F. ($c_p \approx 0.16$ at 900°F.)

$$R_e C_e = \frac{(0.11)(490)r}{(3)(56)} = 0.3208 r_{\text{ft}}$$
$$= 0.01337 D_{\text{in}}$$

Taking the natural log of the transient equation,

$$\ln\left(T_t - T_\infty\right) = \ln\left(\Delta T e^{-t/R_e C_e}\right)$$
$$\ln\left(T_t - T_\infty\right) = \ln \Delta T + \ln\left(e^{-t/R_e C_e}\right)$$
$$\ln\left(T_t - T_\infty\right) = \ln \Delta T - \frac{t}{R_e C_e}$$

$T_t = 250$, $T_\infty = 110$, and $\Delta T = 1800 - 110 = 1690$.

$$\ln(250 - 110) = \ln(1690) - \frac{t}{R_e C_e}$$
$$4.942 = 7.432 - \frac{t}{R_e C_e}$$
$$t = R_e C_e(2.49)$$
$$= (0.01337)D(2.49)$$
$$= 0.0333 D_{\text{in}}$$

(b) The time constant is

$$\frac{c_p \rho L}{h} = \frac{(0.11)(490)\left[\dfrac{D}{(2)(3)(12)}\right]}{56}$$
$$= \boxed{\dfrac{D}{74.81} \text{ hr}}$$

4. Using steam tables,

$$h_1 = 167.99 \text{ BTU/lbm}$$
$$h_2 = 364.17 \text{ BTU/lbm}$$
$$h_3 = 1201.0 \text{ BTU/lbm}$$
$$h_4 = 374.97 \text{ BTU/lbm}$$

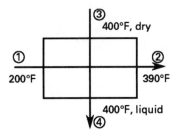

The heat transfer is

$$q = m\Delta h = (500,000)(364.17 - 167.99)$$
$$= 9.809 \times 10^7 \text{ BTU/hr}$$

The flow area per pipe is

$$A = \frac{\pi}{4}D^2 = \frac{\left(\dfrac{\pi}{4}\right)[0.875 - (2 \times 0.0625)]^2}{144}$$
$$= 0.003068 \text{ ft}^2$$

At 390°F, the water volume is largest. From the steam tables,

$$\rho_2 = \frac{1}{0.01850} = 54.05$$

The water volume is

$$Q = \frac{m}{\rho} = \frac{500,000}{54.05} = 9250.7 \text{ ft}^3/\text{hr}$$

The required number of tubes is

$$n_{\text{tube}} = \frac{Q}{A_{\text{tube}} \text{v}} = \frac{9250.7}{(0.003068)(5)(3600)}$$
$$= 167.5 \quad [\text{say } 168]$$

(Normally, 20% extra tubes would be added.)

$$q = F_c U A \Delta T_m$$

Calculate ΔT_m as if in counterflow.

$$\Delta T_A = 400 - 390 = 10°F$$
$$\Delta T_B = 400 - 200 = 200°F$$
$$\Delta T_m = \frac{10 - 200}{\ln\left(\frac{10}{200}\right)} = 63.4°F$$

Even though this is a 2-pass heat exchanger, F_c is not required because one of the fluids has a constant temperature.

The outside area of 1 ft of the entire exchanger (168 tubes) is

$$A = \frac{(168)(1)\pi(0.875)}{12} = 38.48 \text{ ft}^2$$

The required area is

$$A = \frac{q}{U\Delta T_m} = \frac{9.809 \times 10^7}{(700)(63.4)} = 2210 \text{ ft}^2$$

The effective tube length is

$$L = \frac{2210}{38.48} = \boxed{57.4 \text{ ft}}$$

5.
$$h_{probe} = \frac{(0.024)(3480)^{0.8}}{\left(\frac{0.5}{12}\right)^{0.4}}$$
$$= 58.3 \text{ BTU/hr-ft}^2\text{-°R}$$
$$T_{walls} = 600°F = 1060°R$$

The thermocouple gains heat by radiation from the walls. It also loses heat by convection to the gas. Neglect conduction and the insignificant kinetic energy.

$$E_{loss,convection} = E_{gain,radiation}$$
$$E = \frac{q}{A_{probe}} = h(T_{probe} - T_{gas})$$
$$= \sigma\epsilon(T^4_{walls} - T^4_{probe})$$

(a) 300°F = 760°R.

$$(58.3)(T_{probe} - 760) = (0.1713 \times 10^{-8})(0.8)$$
$$\times [(1060)^4 - (T_{probe})^4]$$
$$58.3T_{probe} - 44,308 = 1730 - (1.37 \times 10^{-9})T^4_{probe}$$
$$(1.37 \times 10^{-9})T^4_{probe}$$
$$+ 58.3T_{probe} = 46,038$$

By trial and error,

$$\boxed{T_{probe} = 781°R}$$

(b) If $T_{probe} = 300°F = 760°R$, then

$$h(T_{probe} - T_{gas}) = \sigma\epsilon(T^4_{walls} - T^4_{probe})$$
$$T_{gas} = 760 - \frac{(0.1713 \times 10^{-8})(0.8) \times [(1060)^4 - (760)^4]}{58.3}$$
$$= \boxed{738.2°R}$$

6. Calculate the duct characteristics. The duct diameter is
$$d = \frac{12 \text{ in}}{12 \frac{\text{in}}{\text{ft}}} = 1.0 \text{ ft}$$

The outside duct area is

$$A_{surface} = \pi dL = \pi(1 \text{ ft})(50 \text{ ft}) = 157.1 \text{ ft}^2$$

The cross-sectional area in flow is

$$A_{flow} = \frac{\pi}{4}d^2 = \left(\frac{\pi}{4}\right)(1 \text{ ft})^2 = 0.7854 \text{ ft}^2$$

The entering air density is

$$\rho = \frac{p}{RT}$$
$$= \frac{\left(14.7 \frac{lbf}{in^2}\right)\left(144 \frac{in^2}{ft^2}\right)}{\left(53.3 \frac{ft\text{-}lbf}{lbm\text{-}°R}\right)(45° + 460°)}$$
$$= 0.07864 \text{ lbm/ft}^3$$

The mass flow rate is

$$\dot{m} = \rho Q = \left(0.07864 \frac{lbm}{ft^3}\right)\left(500 \frac{ft^3}{min}\right)\left(60 \frac{min}{hr}\right)$$
$$= 2359.2 \text{ lbm/hr}$$

The mass flow rate per unit area is

$$G = \frac{\dot{m}}{A_{flow}} = \frac{2359.2 \frac{lbm}{hr}}{(0.7854 \text{ ft}^3)\left(3600 \frac{sec}{hr}\right)}$$
$$= 0.8344 \text{ lbm/ft}^2\text{-sec}$$

Since the duct material and thickness were not given, disregard the thermal resistance of the duct.

Estimate temperatures to get initial film coefficients using the given final air temperature.

$$T_{\text{bulk,air}} = \left(\tfrac{1}{2}\right)(T_{\text{air,in}} + T_{\text{air,out}})$$
$$= \left(\tfrac{1}{2}\right)(45°\text{F} + 50°\text{F})$$
$$= 47.5°\text{F} \quad \text{[estimate]}$$
$$T_{\text{surface,duct}} = 70°\text{F at midlength} \quad \text{[estimate]}$$

The film coefficients are not highly sensitive to small temperature differences, so calculate the film coefficients based on these estimates.

The heat transfer inside is forced convection.

$$C = 0.00351 + 0.000001583 T_{\text{bulk}}$$
$$= 0.00351 + (0.000001583)(47.5°\text{F})$$
$$= 0.003585$$
$$h_i = \frac{C(3600G)^{0.8}}{D^{0.2}}$$
$$= \frac{(0.003585)[(3600)(0.8344)]^{0.8}}{(1 \text{ ft})^{0.2}}$$
$$= 2.17 \text{ BTU/hr-ft}^2\text{-°R}$$

The heat transfer outside is natural convection. Estimate h_o at midlength.

$$T_{\text{film}} = \left(\tfrac{1}{2}\right)(T_{\text{surface}} + T_\infty) = \left(\tfrac{1}{2}\right)(70°\text{F} + 80°\text{F})$$
$$= 75°\text{F}$$

At 75°F,

$$\text{Pr} = 0.72$$
$$\frac{g\beta\rho^2}{\mu^2} \approx 2.27 \times 10^6 \frac{1}{\text{ft}^3\text{-°F}}$$
$$\text{Gr} = L^3 \left(\frac{\rho^2 \beta g}{\mu}\right)(T_\infty - T_{\text{surface}})$$
$$\times (1 \text{ ft})^3 \left[2.27 \times 10^6 \frac{1}{\text{ft}^3\text{-°F}}\right]$$
$$\times (80°\text{F} - 70°\text{F})$$
$$= 2.27 \times 10^7$$
$$\text{PrGr} = (2.27 \times 10^7)(0.72)$$
$$= 1.63 \times 10^7$$

The approximate film coefficient for a horizontal cylinder is

$$h_o = (0.27)\left(\frac{T_\infty - T_s}{L}\right)^{0.25}$$
$$= (0.27)\left(\frac{80°\text{F} - 70°\text{F}}{1 \text{ ft}}\right)^{0.25}$$
$$= 0.48 \text{ BTU/hr-ft}^2\text{-°F}$$

The overall heat transfer coefficient is

$$\frac{1}{U} = \frac{1}{h_o} + \frac{1}{h_i}$$
$$= \frac{1}{2.17 \dfrac{\text{BTU}}{\text{hr-ft}^2\text{-°F}}} + \frac{1}{0.48 \dfrac{\text{BTU}}{\text{hr-ft}^2\text{-°F}}}$$
$$= 2.544 \text{ hr-ft}^2\text{-°F/BTU}$$
$$U = \frac{1}{2.544 \dfrac{\text{hr-ft}^2\text{-°F}}{\text{BTU}}} = 0.393 \text{ BTU/hr-ft}^2\text{-°F}$$

The heat transfer due to convection is

$$q_{\text{convection}} = UA(T_\infty - T_{\text{bulk, air}})$$
$$= \left(0.393 \frac{\text{BTU}}{\text{hr-ft}^2\text{-°F}}\right)$$
$$\times (157.1 \text{ ft}^2)(80°\text{F} - 47.5°\text{F})$$
$$= 2006.6 \text{ BTU/hr}$$

Assume the room and duct "see" each other completely. Then $F_e = \epsilon$ and $F_A = 1.0$. The heat exchange due to radiation is

$$q_{\text{radiation}} = \sigma F_e F_A A_{\text{surface}}[T_\infty^4 - T_{\text{surface}}^4]$$
$$= \left(0.1713 \times 10^{-8} \frac{\text{BTU}}{\text{hr-ft}^2\text{-°R}}\right)$$
$$\times (0.28)(1.0)(157.1 \text{ ft}^2)$$
$$\times [(80 + 460)^4 - (70 + 460)^4]$$
$$= 461.6 \text{ BTU/hr}$$

The total heat transfer to the air is

$$q_{\text{total}} = q_{\text{convection}} + q_{\text{radiation}}$$
$$= 2006.6 \frac{\text{BTU}}{\text{hr}} + 461.6 \frac{\text{BTU}}{\text{hr}}$$
$$= 2468.2 \text{ BTU/hr}$$

At 47.5°F, the specific heat of air is approximately 0.240 BTU/lbm-°F. The leaving air temperature is

$$T_{\text{out}} = T_{\text{in}} + \Delta T = T_{\text{in}} + \frac{q_{\text{total}}}{c_p \dot{m}}$$

$$= 45°\text{F} + \frac{2468.2 \dfrac{\text{BTU}}{\text{hr}}}{\left(0.240 \dfrac{\text{BTU}}{\text{lbm-°F}}\right)\left(2359.2 \dfrac{\text{lbm}}{\text{hr}}\right)}$$

$$= 45°\text{F} + 4.36°\text{F} = \boxed{49.36°\text{F}}$$

This agrees with the engineer's estimate.

7. (a) $D_o = \dfrac{4.25}{12} = 0.354 \text{ ft}$

$A_{\text{pipe}} = (0.354)\pi(35 \text{ ft}) = 38.9 \text{ ft}^2$

The entering density is

$$\rho = \frac{(25)(144)}{(53.3)(460 + 500)} = 0.0704 \text{ lbm/ft}^3$$

$$\dot{m} = (200)(0.0704)(60) = 844.8 \text{ lbm/hr}$$

At low pressures, the air enthalpy is found from air tables.

$$h_1 = 231.06 \text{ BTU/lbm at } 960°\text{R}$$

$$h_2 = 194.25 \text{ BTU/lbm at } 810°\text{R}$$

$$q_{\text{loss}} = \dot{m}\Delta h = (844.8)(231.06 - 194.25)$$

$$= 31{,}097 \text{ BTU/hr}$$

The midpoint pipe surface is at $\left(\frac{1}{2}\right)(500 + 350) = 425°$.

$$U = \frac{q}{A\Delta T} = \frac{31{,}097}{(38.9)(425 - 70)}$$

$$= \boxed{2.25 \text{ BTU/hr-ft}^2\text{-}°\text{F}}$$

(b) Disregard the pipe thermal resistance. Also disregard the inside film, which will not be a large factor compared to the outside film and radiation.

Work with the midpoint pipe temperature of 425°F.

Radiation:

Assume $F_a = 1$.

$$F_e = e \approx 0.9 \quad \begin{bmatrix} \text{for } 500°\text{F enamel} \\ \text{paint of any color} \end{bmatrix}$$

$$E = (0.1713 \times 10^{-8})(0.9)$$

$$\times \left[(425 + 460)^4 - (70 + 460)^4\right]$$

$$= 824.1 \text{ BTU/hr-ft}^2$$

Notice the omission of A which would be divided out in the next step. The radiant heat transfer coefficient is

$$h_r = \frac{E}{\Delta T} = \frac{824.1}{425 - 70} = 2.32 \text{ BTU/hr-ft}^2\text{-}°\text{F}$$

Outside coefficient:

Evaluate the film at the midpoint. The film temperature is

$$\left(\tfrac{1}{2}\right)(425 + 70) = 247.5 \quad [\text{say } 250°\text{F}]$$

$$\text{Pr} = 0.72$$

$$\text{Gr} = (0.354)^3(425 - 70)(0.647 \times 10^6) = 1.02 \times 10^7$$

Since $\text{PrGr} < 10^9$, the simplified coefficient is

$$\overline{h} = (0.27)\left(\frac{425 - 70}{0.354}\right)^{0.25} = 1.52$$

The overall film coefficient is

$$h_t = h_r + \overline{h} = 2.32 + 1.52 = \boxed{3.84}$$

This is considerably more than actual.

(c)

- The internal film resistance was disregarded.
- The emmissivity could be lower due to dirty outside of duct.
- h_r and $\overline{h}$ are not really additive.
- Midpoint calculations should be replaced with integration along the length.
- Conductivity of steel pipe is disregarded.

8. $100 \text{ gpm} = (100)(0.1337)(60)(62.4)$

$$= 50{,}057 \text{ lbm/hr}$$

$$q_1 = mc_p\Delta T = (50{,}057)(1)(140 - 70)$$

$$= 350.4 \times 10^4 \text{ BTU/hr}$$

$$\Delta T_A = 230 - 70 = 160$$

$$\Delta T_B = 230 - 140 = 90$$

$$\Delta T_m = \frac{160 - 90}{\ln\left(\dfrac{160}{90}\right)} = 121.66°\text{F}$$

$$U_1 = \frac{q}{A\Delta T_m} = \frac{350.4 \times 10^4}{(50)(121.66)}$$

$$= 576 \text{ BTU/hr-ft}^2\text{-}°\text{F}$$

$$q_2 = (50{,}057)(1)(122 - 70) = 260.3 \times 10^4$$

After fouling,

$$\Delta T_{B,2} = 230 - 122 = 108$$

$$\Delta T_m = \frac{160 - 108}{\ln\left(\dfrac{160}{108}\right)} = 132.3°\text{F}$$

$$U_2 = \frac{260.3 \times 10^4}{(50)(132.3)} = 393.5$$

$$\frac{1}{U_2} = \frac{1}{U_1} + R_f$$

$$R_f = \frac{1}{393.5} - \frac{1}{576}$$

$$= \boxed{0.000805 \text{ hr-ft}^2\text{-}°\text{F/BTU}}$$

9. (a) Heat is lost from the top and sides by radiation and convection.

$$A_{\text{sides}} = \left[\frac{(1.5)(0.75)}{144}\right]\pi = 0.0245 \text{ ft}^2$$

$$A_{\text{top}} = \left(\frac{\pi}{4}\right)\left(\frac{0.75}{12}\right)^2 = 0.003068 \text{ ft}^2$$

$$A_{\text{total}} = A_{\text{sides}} + A_{\text{top}} = 0.0276 \text{ ft}^2$$

$$q = h_t A_1 (T_1 - T_3)$$

$$= h_{\text{sides}} A_{\text{sides}} (T_s - T_\infty) + h_{\text{top}} A_{\text{top}} (T_s - T_\infty)$$

$$+ \sigma F_e F_a A (T_s - T_\infty)$$

For first approximation of T_s, use $h \approx 1.65$ BTU/hr-ft^2-°F.

$$\dot{q} = (5\,W)\left(3.413\,\frac{\dfrac{\text{BTU}}{\text{hr}}}{W}\right)$$

$$= 17.065 \text{ BTU/hr}$$

$$17.065 = (1.65)(0.0245)(T_s - 535)$$

$$+ (1.65)(0.003068)(T_s - 535)$$

$$+ (0.65)(0.0276)(0.1713 \times 10^{-8})$$

$$\times (T_s^4 - (535)^4)$$

By trial and error, $T_s \approx 750$°R.

$$T_f = \left(\tfrac{1}{2}\right)(535 + 750)$$

$$= 642.5 \text{°R} \qquad [\text{use } 640\text{°R}]$$

At $T_f = 640 - 460 = 180$°F,

$$\text{Pr} = 0.72$$

$$\frac{g\beta\rho^2}{\mu^2} = 1.03 \times 10^6$$

$$\text{Gr} = \left(\frac{g\beta\rho^2}{\mu^2}\right) L^3 (T_s - T_\infty)$$

$$(\text{GrPr})_{\text{sides}} = (1.03 \times 10^6)\left(\frac{1.5}{12}\right)^3$$

$$\times (750 - 535)(0.72)$$

$$= 3.11 \times 10^5$$

$$(\text{GrPr})_{\text{top}} = (1.03 \times 10^6)\left(\frac{0.75}{12}\right)^3$$

$$\times (750 - 535)(0.72)$$

$$= 3.89 \times 10^4$$

$$h_{\text{top}} = (0.27)\left(\frac{750 - 535}{\dfrac{0.75}{12}}\right)^{0.25}$$

$$= 2.068$$

$$h_{\text{sides}} = (0.29)\left(\frac{750 - 535}{\dfrac{1.5}{12}}\right)^{0.25}$$

$$= 1.868$$

$$17.065 = (1.868)(0.0245)(T_s - 535)$$

$$+ (2.07)(0.003068)(T_s - 535)$$

$$+ (0.65)(0.0276)(0.1713 \times 10^{-8})[T_s^4 - (535)^4]$$

By trial and error,

$$\boxed{T_s \approx 736\text{°R} = 276\text{°F}}$$

(b)
$$\frac{q_{\text{convection}}}{q_{\text{total}}} = \frac{\begin{array}{c}(1.868)(0.0245)(736 - 535)\\ + (2.068)(0.003068)(736 - 535)\end{array}}{17.065}$$

$$= \boxed{0.614 \ (61.4\%)}$$

$$\frac{q_{\text{radiation}}}{q_{\text{total}}} = \frac{\begin{array}{c}(0.65)(0.0276)(0.1713 \times 10^{-8})\\ \times [(736)^4 - (535)^4]\end{array}}{17.065}$$

$$= \boxed{0.381 \ (38.1\%)}$$

The small error in T_s keeps the total from equaling exactly 100%.

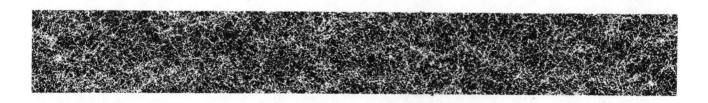

HEATING, VENTILATING, AND AIR CONDITIONING

Warmups

1.

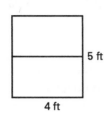

5 ft

4 ft

The total perimeter, including the centerline, is

$$(3)(4) + (2)(5) = 22 \text{ ft}$$

Assume no weatherstripping.

$$B = 80 \ \frac{\text{ft}^3}{\text{hr-ft}}$$

$$Q = \text{BL} = \left(80 \ \frac{\text{ft}^3}{\text{hr-ft}}\right)(22 \text{ ft})$$

$$= \boxed{1760 \text{ ft}^3/\text{hr}}$$

2. The office volume is

$$(60)(95)(10) = 57{,}000 \text{ ft}^3$$

Choose six as the number of air changes per hour. Then,

$$Q = \frac{(57{,}000 \text{ ft}^3)\left(\dfrac{6}{\text{hr}}\right)}{60 \ \dfrac{\text{min}}{\text{hr}}} = \boxed{5700 \text{ ft}^3/\text{min}}$$

From p. 11-6, the preferred ventilation rate is

$$Q = \left(\tfrac{1}{2}\right)(45)(25) + \left(\tfrac{1}{2}\right)(45)(40)$$

$$= \boxed{1463 \text{ ft}^3/\text{min}}$$

3. For the methanol,

Add   2   (because $TLV = 200$)
Add   1   (assuming uniform evolution)
Add   3   (assuming poor booth ventilation)
$k = 6$

$$\text{ventilation} = \frac{(4 \times 10^8)(0.792)(2)(6)}{(32.04)(200)} = 5.9 \times 10^5 \text{ ft}^3/\text{hr}$$

For the methylene chloride,

$$\text{ventilation} = \frac{(4 \times 10^8)(1.336)(2)(6)}{(84.94)(500)} = 1.5 \times 10^5$$

$$\text{total} = (5.9 + 1.5) \times 10^5 = \boxed{7.4 \times 10^5 \text{ ft}^3/\text{hr}}$$

4. A psychrometric chart is needed for this problem.

$$\omega = 79 \text{ gr/lbm dry air}$$

$$= \frac{79 \text{ gr}}{7000 \ \dfrac{\text{gr}}{\text{lbm}}}$$

$$= \boxed{0.01129 \text{ lbm/lbm dry air}}$$

$$h = \boxed{31.5 \text{ BTU/lbm dry air}}$$

$\bar{c}_p$ is gravimetrically weighted. Assume $c_p = 0.241$ BTU/lbm-°F for air.

$$G_{\text{air}} = \frac{1}{1 + 0.01129} = 0.989$$

$$G_{\text{water}} = \frac{0.01129}{1 + 0.01129} = 0.011$$

Using 0.4 as the specific heat of superheated steam,

$$\bar{c}_p = (0.989)(0.241) + (0.011)(0.4)$$

$$= \boxed{0.243 \text{ BTU/lbm-°F}}$$

5. $q = \dfrac{(12{,}000 \text{ W})\left(3413 \ \dfrac{\text{BTU}}{\text{kW-h}}\right)(1.2)}{1000 \ \dfrac{\text{W}}{\text{kW}}}$

$$+ \frac{(12)(0.8)(10 \text{ hp})\left(2545 \ \dfrac{\text{BTU}}{\text{hp-hr}}\right)}{0.9}$$

$$= \boxed{3.206 \times 10^5 \text{ BTU/hr}}$$

6. For Bunker C (No. 6) oil,

$$\text{HV} = \left(\tfrac{1}{2}\right)(151{,}300 + 155{,}900)$$

$$= 153{,}600 \text{ BTU/gal}$$

Assuming $T_i = 70°F$ and $\eta = 0.70$, the fuel consumption is

$$\frac{\left(24 \dfrac{hr}{day}\right)\left(3.5 \times 10^6 \dfrac{BTU}{hr}\right) \times (4772°F\text{-days})}{(70°F - 0°F)(0.70)\left(153,600 \dfrac{BTU}{gal}\right)} = 53,260 \text{ gal}$$

The cost is

$$(53,260 \text{ gal})\left(0.15 \frac{\$}{gal}\right) = \boxed{\$7989}$$

7.

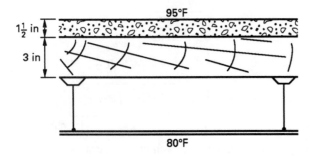

Assume a 15-mph outside wind. Since $h = 6.00$, the film thermal resistance is

$$R_1 = \frac{1}{6} = 0.17$$

For roof insulation used above decking,

$$R_2 = (1.5 \text{ in})\left(\frac{2.78}{in}\right) = 4.17$$

Assume a pine (softwood) deck.

$$R_3 = (3)(1.25) = 3.75$$

$$\epsilon_{wood} = 0.93$$

$$\epsilon_{paper} = 0.95 \quad \text{[used for the acoustical tile]}$$

$$E = \frac{1}{\dfrac{1}{0.93} + \dfrac{1}{0.95} - 1} = 0.89$$

Assuming 4-in separation,

$$C \approx 0.81$$

$$R_4 = \frac{1}{0.81} = 1.23$$

For $\frac{3}{4}$-in acoustical tile,

$$R_5 = 1.78$$

For the inside film, $h = 1.65$.

$$R_6 = \frac{1}{1.65} = 0.61$$

The total resistance is

$$\begin{aligned} R_t &= R_1 + R_2 + R_3 + R_4 + R_5 + R_6 \\ &= 0.17 + 4.17 + 3.75 + 1.23 + 1.78 + 0.61 \\ &= 11.71 \end{aligned}$$

$$U = \frac{1}{R_t} = \frac{1}{11.71}$$

$$= \boxed{0.0854 \text{ BTU/hr-ft}^2\text{-°F}}$$

8. $BF = \left(\dfrac{1}{3}\right)^4 = \boxed{0.0123}$

9. Use the slab edge method.

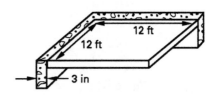

$$P = 12 + 12 = 24 \text{ ft}$$

Choose $F = 0.55$. Assume $T_i = 70°F$.

$$q = (24 \text{ ft})\left(0.55 \frac{BTU}{hr\text{-}ft\text{-}°F}\right)[70 - (-10)]$$

$$= \boxed{1056 \text{ BTU/hr}}$$

10. (a) The total ventilation is

$$Q = \left(60 \frac{\frac{ft^3}{min}}{person}\right)(4500 \text{ people})\left(60 \frac{min}{hr}\right)$$

$$= 1.62 \times 10^7 \text{ ft}^3/\text{hr}$$

Because $T_o = 0°F$ (less than freezing), there will be no moisture in the air. The outside density is

$$\rho = \frac{p}{RT} = \frac{(14.6 \text{ psi})(144)}{(53.3)(460 + 0°F)}$$

$$= 0.08575 \text{ lbm/ft}^3$$

$$\dot{m} = Q\rho = \left(1.62 \times 10^7 \frac{ft^3}{hr}\right)\left(0.08575 \frac{lbm}{ft^3}\right)$$

$$= 1.389 \times 10^6 \text{ lbm/hr}$$

Assume $c_p = 0.241$ BTU/lbm-°F. The increase in air temperature once it enters the auditorium is due to the body heat, less any conductive losses through the walls, etc. The occupant sensible load is 225 BTU/hr-person. (The latent load adds moisture to the air but does not increase temperature.)

$$q_{\text{in from people}} = (225)(4500) = 1.01 \times 10^6 \text{ BTU/hr}$$

The air leaves at 70°F.

$$q = mc_p\Delta T$$
$$1.01 \times 10^6 = (1.389 \times 10^6)(0.241)(70 - T_{\text{in}})$$
$$T_{\text{in}} = \boxed{66.98°\text{F}}$$

(b) The coil heat needed is

$$q = mc_p\Delta T$$
$$= (1.389 \times 10^6)(0.241)(66.98 - 0)$$
$$= \boxed{2.24 \times 10^7 \text{ BTU/hr}}$$

(c) $\boxed{\begin{array}{l}\text{Since } 2.24 \times 10^7 > 1.25 \times 10^7, \\ \text{the furnace is too small.}\end{array}}$

Concentrates

1.

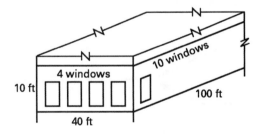

The total exposed (glass and wall) area is

$$(10)(100 + 100 + 40) = 2400 \text{ ft}^2$$

The window area is

$$(10 \text{ windows} + 10 \text{ windows} + 4 \text{ windows})\left[(4 \text{ ft})(6 \text{ ft})\right]$$
$$= 576 \text{ ft}^2$$

The inside and outside film coefficients are 1.65 and 6.00 BTU/hr-ft^2-°F, respectively.

$$U_{\text{wall}} = \cfrac{1}{\cfrac{1}{1.65} + \cfrac{1}{0.2} + \cfrac{1}{6.00}} = 0.173$$

The heat loss for the walls is

$$q_1 = UA\Delta T = (0.173)(2400 - 576)[75 - (-10)]$$
$$= 26{,}820 \text{ BTU/hr}$$

For the windows assuming a $\frac{1}{4}$-in air space,

$$U_{\text{windows}} = \frac{1}{1.63} = 0.61$$

The heat loss for the windows is

$$q_2 = (0.61)(576)[75 - (-10)]$$
$$= 29{,}870 \text{ BTU/hr}$$

For the sake of completeness, calculate the air infiltration. Assume double-hung, unlocked windows.

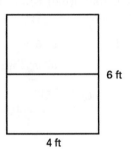

$$B = 32 \text{ cfh/ft}$$
$$L = 4 + 4 + 4 + 6 + 6 = 24 \text{ ft}$$
$$Q = (14 \text{ windows})(32)(24)$$
$$= 10{,}750 \text{ ft}^3/\text{hr}$$

(Use windows on two sides for three exposed walls.)

$$q_3 = (0.018)(10{,}750)[75 - (-10)] = 16{,}450$$
$$q_{\text{total}} = q_1 + q_2 + q_3$$
$$= 26{,}820 + 29{,}870 + 16{,}450$$
$$= \boxed{73{,}140 \text{ BTU/hr}}$$

2. Locate points A and B on the psychrometric chart.

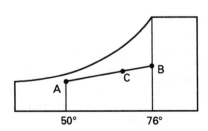

The length of line A-B is 6.6 cm. Reading from the chart,

$$v_A = 13.0 \text{ ft}^3/\text{lbm}$$

$$v_B = 13.69 \text{ ft}^3/\text{lbm}$$

$$\rho_A = \frac{1}{13.0} = 0.0769 \text{ lbm/ft}^3$$

$$\rho_B = \frac{1}{13.69} = 0.073 \text{ lbm/ft}^3$$

$$m_A = \rho Q_A = (0.0769)(1000)$$

$$= 76.9 \text{ lbm}$$

$$m_B = (0.073)(1500) = 109.5 \text{ lbm}$$

The gravimetric fraction of flow A is

$$\frac{76.9}{76.9 + 109.5} = 0.41$$

Point C is located $(0.41)(6.6) = 2.7$ cm from point B. This determines

$$T_{\text{dry bulb}} = 65.5°\text{F}$$

$$\omega = 57 \text{ gr/lbm}$$

$$T_{\text{dp}} = \boxed{51°\text{F}}$$

3. $$\text{BF} = 1 - 0.70 = \boxed{0.30}$$

$$0.70 = \frac{60 - T_{\text{out}}}{60 - 45}$$

$$T_{\text{out}} = \boxed{49.5°\text{F}}$$

4. From the psychrometric chart,

At 1,

$$\omega_1 = 99 \text{ gr/lbm}$$

$$h_1 = 38.3 \text{ BTU/lbm}$$

$$v_1 = 14.29 \text{ ft}^3/\text{lbm}$$

At 2,

$$\omega_2 = 165.8$$

$$h_2 = 46.4$$

The moisture added is

$$\frac{165.8 \dfrac{\text{gr}}{\text{lbm air}} - 99 \dfrac{\text{gr}}{\text{lbm air}}}{\left(7000 \dfrac{\text{gr}}{\text{lbm water}}\right)\left(14.29 \dfrac{\text{ft}^3}{\text{lbm air}}\right)}$$

$$= \boxed{6.68 \times 10^{-4} \text{ lbm/ft}^3}$$

The enthalpy change is

$$\frac{46.4 - 38.3}{14.29} = \boxed{0.567 \text{ BTU/ft}^3 \text{ air}}$$

5. $$\text{SHR} = \frac{200,000}{200,000 + 50,000} = 0.8$$

Draw the condition line.

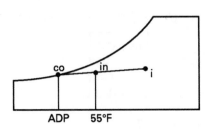

$$\text{ADP} = 50.8°\text{F}$$

$$T_w = \text{ADP}$$

Choose 20°F as a temperature gain.

$$T_{\text{db,in}} = 75 - 20 = 55°\text{F}$$

$$Q_{\text{in}} = \frac{(55.3)(200,000)}{(60)(75 - 55)} = 9217 \text{ ft}^3/\text{min}$$

This is a mixing problem.

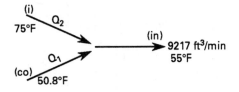

The lines (co-i) and (in-i) have lengths of 6.15 cm and 5.05 cm, respectively.

$$\frac{5.05}{6.15} = 0.821$$

$$Q_1 = (0.821)(9217) = \boxed{7567 \text{ ft}^3/\text{min}}$$

6. Assume that the heating load is based on $T_{\text{id}} = 70°\text{F}$. The heat loss per degree is

$$\frac{q}{\Delta T} = \frac{650,000}{70 - 0}$$

$$= 9286 \text{ BTU/hr-°F}$$

The average outdoor temperature during the heating season is found from

$$5252 = (245)(65 - \overline{\overline{T}}_o)$$

$$\overline{\overline{T}}_o = 43.56°F$$

The period between 8:30 a.m. and 5:30 p.m. is 9 hr. The total winter heat loss is

$$(245)(9286)\big[(9)(70 - 43.56)$$
$$+ (15)(50 - 43.56)\big] = 7.61 \times 10^8 \text{ BTU}$$

The fuel consumption is

$$\frac{7.61 \times 10^8}{(13{,}000)(0.70)} = \boxed{83{,}630 \text{ lbm/yr}}$$

7. This is a mixing problem. The process cannot be represented by a constant enthalpy line because $T_{\text{water}} < T_{\text{wb, incoming}}$.

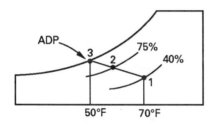

At 1,

$$h_1 = 23.6 \text{ BTU/lbm}$$

$$\omega_1 = 44 \text{ gr/lbm}$$

$$v_1 = 13.48 \text{ ft}^3/\text{lbm}$$

The mass of incoming air is

$$\dot{m}_1 = \frac{1800 \dfrac{\text{ft}^3}{\text{min}}}{13.48 \dfrac{\text{ft}^3}{\text{lbm}}} = 133.53 \text{ lbm/min}$$

At 2,

$$h_2 = 21.4 \text{ BTU/lbm}$$

$$\omega_2 = 51 \text{ gr/lbm}$$

$$T_{\text{db}, 2} = 56.3°F$$

$$T_{\text{wb}} = 51.9°F$$

The added water is

$$\frac{\left(133.53 \dfrac{\text{lbm-air}}{\text{min}}\right)\left(51 \dfrac{\text{gr water}}{\text{lbm air}} - 44 \dfrac{\text{gr water}}{\text{lbm air}}\right)}{7000 \dfrac{\text{gr water}}{\text{lbm water}}}$$

$$= \boxed{0.134 \text{ lbm/min}}$$

8. Use the energy and weight balances. From problem 7,

$$\omega_1 = \frac{44 \dfrac{\text{gr}}{\text{lbm}}}{7000 \dfrac{\text{gr}}{\text{lbm}}} = 0.006286 \text{ lbm/lbm}$$

$$h_1 = 23.6 \text{ BTU/lbm}$$

$$\dot{m} = 133.53 \text{ lbm/min}$$

For 1 atm steam, $h = 1150.4$ BTU/lbm.

The mass balance is

$$(133.53)(1 + 0.006286) + m_{\text{steam}} = (133.53)\left(1 + \frac{\omega_2}{7000}\right)$$

The energy balance is

$$(133.53)(23.6) + m_{\text{steam}}(1150.4) = (133.53)h_2$$

Since no single relationship exists between ω_2, m_{steam}, and h_2, a trial-and-error solution is required. Once m_{steam} is selected, ω_2 and h_2 can be found from the previous two equations.

$$\omega_2 = (52.42)m_{\text{steam}} + 44$$

$$h_2 = (8.615)m_{\text{steam}} + 23.6$$

If ω_2 and h_2 are known, the relative humidity can be determined. If the relative humidity is 75%, then ω_{steam} was chosen correctly. The trial-and-error iterations are

m_{steam}	ω_2	h_2	RH
0.3	59.7	26.2	53%
0.5	70.2	27.9	61%
0.7	80.7	29.6	72%
0.8	86	30.5	75%

$$\boxed{\begin{aligned} \omega_2 &= 86 \text{ gr/lbm} \\ h_2 &= 30.5 \text{ BTU/lbm} \\ T_{\text{db}} &= 71.1°F \\ T_{\text{wb}} &= 65.6°F \end{aligned}}$$

9. For the incoming air,

$$v_1 = 13.37 \text{ ft}^3/\text{lbm}$$

$$\omega_1 = 51 \text{ gr/lbm}$$

With sensible heating as the limiting factor,

$$Q = \frac{500{,}000}{(0.018)(75 - 65)}$$

$$= 2.777 \times 10^6 \text{ ft}^3/\text{hr}$$

$$\dot{m} = \frac{2.777 \times 10^6}{13.37}$$

$$= 2.078 \times 10^5 \text{ lbm/hr}$$

Assume that this air absorbs all the moisture. Then the final humidity will be

$$\omega_2 = 51 + \frac{\left(175 \ \frac{\text{lbm}}{\text{hr}}\right)\left(13.37 \ \frac{\text{ft}^3}{\text{lbm}}\right)\left(7000 \ \frac{\text{gr}}{\text{lbm}}\right)}{2.777 \times 10^6 \ \frac{\text{ft}^3}{\text{hr}}}$$

$$= 56.9 \ \text{gr/lbm}$$

The final conditions are

$$T_{\text{db}} = 75°\text{F}$$
$$\omega_2 = 56.9 \ \text{gr/lbm}$$
$$\text{RH} = 43\%$$

This is below RH = 60%.

10. Locate points (out) and (co) on the psychrometric chart.

$$v_{\text{out}} = 13.94 \ \text{ft}^3/\text{lbm}$$
$$h_{\text{out}} = 36.2 \ \text{BTU/lbm}$$

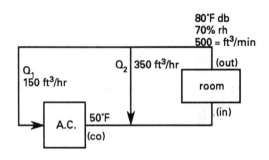

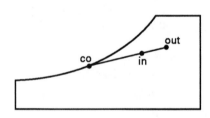

$$h_{\text{co}} = 20.2 \ \text{BTU/lbm} \quad \text{[saturated]}$$

$$\dot{m}_1 = \frac{150 \ \frac{\text{ft}^3}{\text{min}}}{13.94 \ \frac{\text{ft}^3}{\text{lbm}}}$$

$$= 10.76 \ \text{lbm/min}$$

$$\dot{m}_2 = \frac{350}{13.94} = 25.11 \ \text{lbm/min}$$

The percent bypass is

$$x = \frac{25.11}{25.11 + 10.76} = 0.70 \ \ (70\%)$$

On the chart, the length of line (co-out) is 8.5 cm. Point (in) is located $(0.70)(8.5) = 6$ cm from point (co).

(a) At that point,

$$\boxed{\begin{aligned} T_{\text{db, in}} &= 71.2°\text{F} \\ \omega_{\text{in}} &= 92 \ \text{gr/lbm} \\ \text{RH}_{\text{in}} &= 80\% \end{aligned}}$$

(b) The conditioner capacity is

$$\frac{\left(10.76 \ \frac{\text{lbm}}{\text{min}}\right)\left(36.2 \ \frac{\text{BTU}}{\text{lbm}} - 20.2 \ \frac{\text{BTU}}{\text{lbm}}\right)}{200 \ \frac{\text{BTU}}{\text{min-ton}}} = \boxed{0.86 \ \text{ton}}$$

Timed

1. (a)

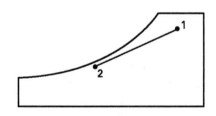

At 1,

$$\boxed{\begin{aligned} h_1 &= 50.7 \ \text{BTU/lbm} \\ v_i &= 14.55 \ \text{ft}^3/\text{lbm} \\ \omega_i &= 177 \ \text{gr/lbm} \end{aligned}}$$

At 2,

$$\boxed{\begin{aligned} h_2 &= 25.8 \ \text{BTU/lbm} \\ \omega_2 &= 73 \ \text{gr/lbm} \end{aligned}}$$

(b) The air mass is

$$\frac{5000 \ \frac{\text{ft}^3}{\text{min}}}{14.55 \ \frac{\text{ft}^3}{\text{lbm}}} = 343.6 \ \text{lbm/min}$$

The water decrease is

$$\frac{\left(343.6 \ \frac{\text{lbm air}}{\text{min}}\right)\left(177 \ \frac{\text{gr}}{\text{lbm air}} - 73 \ \frac{\text{gr}}{\text{lbm air}}\right)}{7000 \ \frac{\text{gr}}{\text{lbm water}}}$$

$$= \boxed{5.1 \ \text{lbm/min}}$$

(c) $\left(343.6 \ \dfrac{\text{lbm}}{\text{min}}\right)\left(50.7 \ \dfrac{\text{BTU}}{\text{lbm}} - 25.8 \ \dfrac{\text{BTU}}{\text{lbm}}\right)$

$$= \boxed{8556 \ \text{BTU/min}}$$

(d) It is not clear from the problem whether this is a wet or dry cycle. Assume that it is a wet cycle since the problem says "saturated" at 100°F.

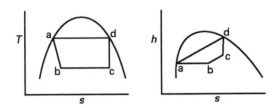

(e) At a,

   $T = 100°\text{F}$
   $p = 131.6 \ \text{psia}$
   $h = 31.16 \ \text{BTU/lbm}$
   $s = 0.06316 \ \text{BTU/lbm-}°\text{F}$
   $v = 0.0127 \ \text{ft}^3/\text{lbm}$

At b,

   $T = 50°\text{F}$
   $p = 61.39 \ \text{psia}$
   $h = h_a = 31.16 \ \text{BTU/lbm}$
   $x = \dfrac{31.16 - 19.27}{64.51} = 0.184$
   $s = 0.04126 + (0.184)(0.12659)$
   $\ \ = 0.06455 \ \text{BTU/lbm-}°\text{F}$
   $v = 0.0118 + (0.184)(0.673 - 0.0118)$
   $\ \ = 0.1335 \ \text{ft}^3/\text{lbm}$

At d,

   $T = 100°\text{F}$
   $p = 131.6 \ \text{psia}$
   $h = 88.62 \ \text{BTU/lbm}$
   $s = 0.16584 \ \text{BTU/lbm-}°\text{F}$
   $v = 0.319 \ \text{ft}^3/\text{lbm}$

At c,

   $T = 50°\text{F}$
   $p = 61.39 \ \text{psia}$
   $s = s_d = 0.16584 \ \text{BTU/lbm-}°\text{F}$
   $x = \dfrac{0.16584 - 0.04126}{0.12659} = 0.984$
   $h = 19.27 + (0.984)(64.57) = 82.80$
   $v = 0.0118 + (0.984)(0.673 - 0.0118)$
   $\ \ = 0.662 \ \text{ft}^3/\text{lbm}$

2. The heat loss per °F is

$$\frac{q}{\Delta T} = \frac{200{,}000}{75 - 0} = 2667 \ \text{BTU/hr-}°\text{F}$$

The average outdoor temperature during the heating season is found using the following analysis.

$$DD = 4200 = \sum_{i=1}^{210}(65 - \overline{T}_o)$$

$$= (210)(65 - \overline{\overline{T}})$$

$$\overline{\overline{T}} = 45°\text{F}$$

$\overline{T}$ and $\overline{\overline{T}}$ are not the same. $\overline{\overline{T}}$ is the average temperature over all hours in the entire 210-day heating season.

The original heat loss during the heating season was

$$q = \left(24 \ \frac{\text{hr}}{\text{day}}\right)(210 \ \text{days})\left(2667 \ \frac{\text{BTU}}{\text{hr-}°\text{F}}\right)$$
$$\times (75°\text{F} - 45°\text{F})$$
$$= 4.03 \times 10^8 \ \text{BTU}$$

The reduced heat loss is

$$q = (24)(210)(2667)(68 - 45)$$
$$= 3.09 \times 10^8 \ \text{BTU}$$

The reduction is

$$\frac{4.03 - 3.09}{4.03} = \boxed{0.233 \ (23.3\%)}$$

3. This is a standard bypass problem.

The indoor conditions are

$$T_i = 75°\text{F}$$
$$\phi_i = 50\%$$

The outdoor conditions are

$$T_{o,\text{db}} = 90°\text{F}$$
$$T_{o,\text{wb}} = 76°\text{F}$$

The ventilation is 2000 ft³/min.

The loads are

$$q_s = 200{,}000 \ \text{BTU/hr}$$
$$q_l = 450{,}000 \ \text{gr/hr}$$

To find the sensible heat ratio, q_l must be expressed in BTU/hr.

For the room conditions, $\omega \approx 0.0095$. (Actually, it should be a little less since moisture is removed between T_{in} and T_i.) Assume $p = 14.7$ psia.

$$0.0095 = \frac{0.622p_w}{14.7 - p_w}$$

$$\approx \frac{0.622p_w}{14.7}$$

$$p_w = 0.22 \text{ psia}$$

For $p = 0.22$ psia, from the steam tables,

$$h_{fg} = 1060 \text{ BTU/lbm}$$

Each pound of water evaporated requires approximately 1060 BTU.

$$q_1 = \frac{\left(450,000 \ \frac{\text{gr}}{\text{hr}}\right)\left(1060 \ \frac{\text{BTU}}{\text{lbm}}\right)}{7000 \ \frac{\text{gr}}{\text{lbm}}}$$

$$= 68,140 \text{ BTU/hr}$$

$$\text{SHR} = \frac{200,000}{200,000 + 68,140} = 0.75$$

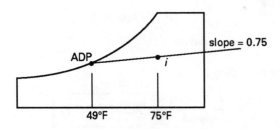

Using T_i and ϕ_i, locate point i on the psychrometric chart, and draw the condition line with slope = 0.75 through it.

The left-hand intersection shows ADP= 49°F. Since the air leaves the conditioner saturated,

$$\text{BF}_{\text{coil}} = 0$$

$$T_{co} = \boxed{49°F}$$

Calculate the air flow through the room.

$$Q_{in} = \frac{200,000}{\left(\dfrac{60}{55.3}\right)(75 - 58)}$$

$$= \boxed{10,840 \text{ ft}^3/\text{min}}$$

$$Q_{in} = (60)(10,840)$$

$$= 650,400 \text{ ft}^3/\text{hr}$$

Since $T_{in} = 58°F$ is given, locate this dry bulb temperature on the condition line.

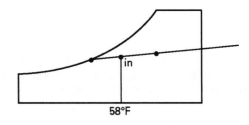

Reading from the chart,

$$\boxed{\omega_{in} = 56.5 \text{ gr/lbm}}$$

4. Although high temperature psychrometric charts exist, it is not necessary to use them. At $T_1 = 200°F$ and 100% relative humidity,

$$p_w = 11.526 \text{ psia}$$

Since $p_t = 25$ psia,

$$p_a = 25 - 11.526 = 13.474 \text{ psia}$$

The specific gas constant of water is 85.8.

The mass of water vapor is constant.

$$m_w = \frac{pV}{RT} = \frac{(11.526)(144)(1500)}{(85.8)(200 + 460)}$$

$$= 43.96 \text{ lbm}$$

The mass of the air is also constant.

$$m_a = \frac{(13.474)(144)(1500)}{(53.3)(200 + 460)}$$

$$= 82.73 \text{ lbm}$$

$$\omega = \frac{43.96}{82.73} = 0.531$$

The masses of air and water do not change, so the mole fractions and partial pressure ratios also do not change. $p_{w,2} = 11.526$. At $T_2 = 400°F$,

$$p_{w,\text{sat}} = 247.31 \text{ psi}$$

$$\phi_2 = \frac{11.526}{247.31} = \boxed{0.0466}$$

ω is unchanged.

$$\omega_2 = \boxed{0.531}$$

The energy required must be obtained in two parts.

CHAPTER 11   HEATING, VENTILATING, AND AIR CONDITIONING

Air:

$$h_1 = 157.92 \text{ BTU/lbm} \quad \text{[at } 660°\text{R]}$$

$$h_2 = 206.46 \text{ BTU/lbm} \quad \text{[at } 860°\text{R]}$$

$$q_1 = 206.46 - 157.92$$

$$= 48.54 \text{ BTU/lbm-dry air}$$

Water: On the Mollier diagram, locate point 1 ($T = 200°$F, $p = 11.526$).

$$h_1 = 1146 \text{ BTU/lbm} \quad \text{[appears to be saturated]}$$

Follow a constant pressure curve up to $400°$F.

$$h_2 = 1240 \text{ BTU/lbm}$$

$$q_2 = \frac{(43.96)(1240 - 1146)}{82.73}$$

$$= 49.95 \text{ BTU/lbm-dry air}$$

$$q_{\text{total}} = 48.54 + 49.95$$

$$\boxed{= 98.49 \text{ BTU/lbm-dry air}}$$

The dew point is the temperature at which the water starts to condense out in a constant pressure process. Following the constant pressure line back to the saturation line,

$$\boxed{T_{\text{dp}} = 200°\text{F}}$$

5. Annual savings accrue during the 21-week heating season. Assume negligible crack losses compared to the volume of forced ventilation air. Also neglect humidification differences since moisture content is not given (outside air temperature and moisture).

With air at $70°$F, $c_p \approx 0.240$ BTU/lbm-°F, and $\rho \approx 0.075$ lbm/ft^3.

In one air change,

$$q_{\text{air}} = mc_p\Delta T = (801,000 \text{ ft}^3)\left(0.075 \frac{\text{lbm}}{\text{ft}^3}\right)$$

$$\times \left(0.240 \frac{\text{BTU}}{\text{lbm}°\text{F}}\right)(12°\text{F})$$

$$= 173,016 \text{ BTU}$$

Then, on an hourly basis,

$$q_{\text{air/hr}} = \left(\frac{173,016 \text{ BTU}}{\text{air change}}\right)\left(\frac{\frac{1}{2} \text{ air change}}{\text{hour}}\right)$$

$$= 86,508 \text{ BTU/hr}$$

For the roof, floor, and walls,

$$q = UA\Delta T$$

$$= \big[(0.15)(11,040) + (1.13)(2760) + (0.05)(26,700)$$

$$+ (1.5)(690)\big](12°\text{F})$$

$$= 85,738 \text{ BTU/hr}$$

Unoccupied time:

$$t = \left[(14)(5)\frac{\text{hr}}{\text{wk}} + 48\frac{\text{hr}}{\text{wk}}\right](21 \text{ weeks})$$

$$= 2478 \text{ hr/yr}$$

$$q_{\text{lost/year}} = \left(86,508 \frac{\text{BTU}}{\text{hr}} + 85,738 \frac{\text{BTU}}{\text{hr}}\right)$$

$$\times \left(2478 \frac{\text{hr}}{\text{yr}}\right)$$

$$= \frac{4.27 \times 10^8 \text{ BTU}}{100,000 \frac{\text{BTU}}{\text{therm}}}$$

$$= 4270 \text{ therms}$$

$$q_{\text{required}} = \frac{4270 \text{ therms}}{0.75} = 5693 \text{ therms}$$

$$\text{savings} = (5693 \text{ therms})\left(\frac{\$0.25}{\text{therm}}\right)$$

$$\boxed{= \$1423 \text{ per year}}$$

6. (a).

$$800°\text{F} = 1260°\text{R}$$

$$350°\text{F} = 810°\text{R}$$

From the air tables,

$$h_{1,\text{air},800°\text{F}} = 306.65 \text{ BTU/lbm}$$

$$h_{2,\text{air},350°\text{F}} = 194.25 \text{ BTU/lbm}$$

From the steam tables at $80°$F or from a Mollier diagram,

$$h_{1,\text{water}} = 48.02 \text{ BTU/lbm}$$

$$h_{2,\text{water}} = (0.444)(350) + 1061$$

$$= 1216.4 \text{ BTU/lbm}$$

$$m_w = \frac{m_{\text{air}}(\Delta h_{\text{air}})}{\Delta h_w}$$

$$= \frac{\left(410 \frac{\text{lbm}}{\text{hr}}\right)(194.25 - 306.65)}{48.02 - 1216.4} \quad \begin{bmatrix} \text{use } \Delta h = c_p\Delta T \\ \text{only if } c_p \text{ is} \\ \text{assumed} \\ \text{constant} \end{bmatrix}$$

$$\boxed{= 39 \text{ lbm/hr water}}$$

(b)

$$\phi = \frac{p_w}{p_{ws}}$$

$$p_{\text{sat},350°} = 134.63 \text{ psia}$$

PROFESSIONAL PUBLICATIONS, INC. ● Belmont, CA

The partial pressure is proportional to the mole fraction.

$$\text{no. moles water} = \frac{40}{18} = 2.22$$

$$\text{no. moles air} = \frac{410}{29} = 14.14$$

$$x_w = \frac{2.22}{2.22 + 14.14} = 0.136$$

$$p_w = (0.136)(100) = 13.6 \text{ psia}$$

$$\phi = \frac{13.6}{134.63} = \boxed{0.101 \ (10.1\%)}$$

7. (a) The cooled water flow rate is

$$\frac{q}{c_p \Delta T} = \frac{1 \times 10^6 \ \frac{\text{BTU}}{\text{hr}}}{\left(1 \ \frac{\text{BTU}}{\text{lbm-}°\text{F}}\right)(120°\text{F} - 110°\text{F})}$$

$$= 1 \times 10^5 \text{ lbm/hr}$$

For air in, using the psychrometric diagram,

$$h \approx 43.0 \text{ BTU/lbm}$$

For the air out, the psychrometric diagram (offscale) cannot be used, so use equations. From a saturated steam table at 100°F, $p_{ws} = 0.9492$ psia.

$$\omega_{\text{air out}} = \frac{(0.622)(0.82)(0.9492)}{14.696 - (0.82)(0.9492)}$$

$$= 0.035$$

Then, for the air out,

$$h_2 = (0.241)(100) + (0.035)[(0.444)(100) + 1061]$$

$$= 62.79 \text{ BTU/lbm}$$

Since $q_{\text{removed}} = \dot{m}_{\text{air}}(h_2 - h_1)$,

$$\dot{m}_{\text{air}} = \frac{1 \times 10^6 \ \frac{\text{BTU}}{\text{hr}}}{62.79 \ \frac{\text{BTU}}{\text{lbm}} - 43.0 \ \frac{\text{BTU}}{\text{lbm}}}$$

$$= \boxed{5.05 \times 10^4 \text{ lbm/hr air}}$$

(b) The humidity ratio can be found using the Mollier diagram or as follows.

Conservation of water vapor:

$$\omega_1 \dot{m}_{\text{air},1} + \dot{m}_{\text{make-up}} = \omega_2 \dot{m}_{\text{air},2}$$

$$\dot{m}_{\text{make-up}} = \dot{m}_{\text{air}}(\omega_2 - \omega_1)$$

From a steam table at 91°F, $p_{ws} = 0.7233$ psia. Then,

$$p_w = (0.60)(0.7233) = 0.434 \text{ psia}$$

$$\omega_3 = \frac{(0.622)(0.434)}{14.696 - 0.434} = 0.019$$

Make-up water:

$$\dot{m}_m = (5.05 \times 10^4)(0.035 - 0.019)$$

$$= \boxed{808 \text{ lbm/hr water}}$$

8. First, determine thermal resistances.

walls	R	notes
4-in brick facing	0.44	
3-in concrete block	0.40	
1-in mineral wool	3.33	
2-in furring	0.93	(indicates 2-in air space)

For 2-in air space, $e_{\text{mineral wool}}$ should be roughly $e_{\text{iron oxide}} = 0.96$. Likewise, $e_{\text{drywall gypsum}}$ should be about $e_{\text{white paper}} = 0.95$.

$$\frac{1}{E} = \frac{1}{0.96} + \frac{1}{0.95} - 1 = 1.09$$

$$E = 0.91$$

For a vertical air space, $C = 1.08$.

$$R = \frac{1}{C} = \frac{1}{1.08}$$

$$= 0.93$$

walls	R	notes
$\frac{3}{8}$-in drywall gypsum	0.34	$\left(\frac{\frac{3}{8}}{\frac{1}{2}}\right)(0.45) = 0.3375$
$\frac{1}{2}$-in plaster	0.09	
surface outside	0.25	
surface inside	0.61	still air, $h \sim 1.65$

roof	R	determined from
4-in concrete	0.32	sand aggregate
2-in insulation	5.56	
felt	0.06	
1-in air gap	0.87	
ceiling tile	1.19	$\frac{1}{2}$-in acoustical tile
surface outside	0.25	summer conditions, downward heat flow, 7.5 mph winds
surface inside	0.92	

windows	R	determined from
$\frac{1}{4}$-in thick, single glazing (includes film coefficient)	0.66	without blinds, $R = 0.88$, $R = (0.75)(0.88) = 0.66$

Next, determine the heat fluxes.

Walls: Using Eq. 10.7,

$$U = \frac{1}{\begin{array}{c}0.25 + 0.44 + 0.40 + 3.33 \\ + 0.93 + 0.34 + 0.09 + 0.61\end{array}}$$

$$= 0.16$$

For 4-in brick facing at 4 p.m.,

$$q_{\text{walls}} = (0.16)\big[(1600)(17) + (1400)(28) + (1500)(20)$$

$$+ (1400)(20)\big]$$

$$= 19,900 \text{ BTU/hr}$$

Roof:

$$U = \frac{1}{0.25 + 0.32 + 5.56 + 0.06 + 0.87 + 1.19 + 0.92}$$

$$= 0.11$$

For heavy 4-in concrete roof at 4 p.m.,

$$q_{\text{roof}} = (0.11)(6000)(74) = 48,840 \text{ BTU/hr}$$

Windows: Since all windows face east and the calculation is for 4 p.m., assume windows are not in direct sunlight.

$$q_{\text{windows}} = \left(\frac{1}{0.66}\right)(100)(95 - 78) = 2576 \text{ BTU/hr}$$

Sensible transmission load, then, is

$$q_{\text{total}} = 19,900 + 48,840 + 2576$$

$$\boxed{= 71,320 \text{ BTU/hr}}$$

This may not be the peak cooling load; the outdoor temperature at other times must be known. See the chart below. The peak cooling load appears between 4:00 and 6:00 p.m.

	q (BTU/hr)				
	12 p.m.	2 p.m.	4 p.m.	6 p.m.	8 p.m.
walls	13,984	18,112	19,900	24,512	23,328
roof	30,084	42,510	48,840	44,472	32,046
windows	1818 (90°F) or 2576 (95°F)				
total $\left(\begin{array}{c}90°F \\ \text{outside}\end{array}\right)$	45,886	62,440	70,576	70,802	57,192
total $\left(\begin{array}{c}95°F \\ \text{inside}\end{array}\right)$	46,644	63,198	71,320	71,560	57,950

STATICS

Warmups

1.

$$T = Fr = \frac{(3\ \text{lbf})(3\ \text{in})}{12\dfrac{\text{in}}{\text{ft}}}$$

$$= \boxed{0.75\ \text{ft-lbf}}$$

2.

$$E_{\text{steel}} = 30 \times 10^6\ \text{psi}$$

$$E_{\text{alum}} = 10 \times 10^6\ \text{psi}$$

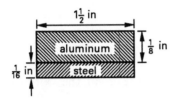

$$n = \frac{30 \times 10^6}{10 \times 10^6} = 3$$

The equivalent, all-aluminum cross section is

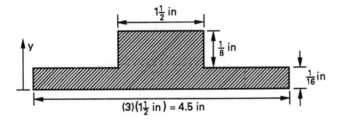

$$A_1 = \left(\frac{1}{16}\right)(4.5)$$

$$= 0.28125\ \text{in}^2$$

$$A_2 = \left(\frac{1}{8}\right)(1.5)$$

$$= 0.1875\ \text{in}^2$$

$$y_1 = \frac{\frac{1}{16}}{2}$$

$$= 0.03125\ \text{in}$$

$$y_2 = \frac{1}{16} + \frac{\frac{1}{8}}{2}$$

$$= 0.125\ \text{in}$$

$$\bar{y} = \frac{(0.28125)(0.03125) + (0.1875)(0.125)}{0.28125 + 0.1875}$$

$$= 0.06875\ \text{in}$$

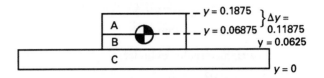

The centroidal moments of inertia for sections A and B are

$$I_A = \frac{bh^3}{3} = \frac{(1.5)(0.11875)^3}{3} = 8.373 \times 10^{-4}\ \text{in}^4$$

$$I_B = \frac{(1.5)(0.00625)^3}{3} = 1.221 \times 10^{-7}\ \text{in}^4$$

The moment of inertia of section C about its own centroid is

$$I_{ox} = \frac{bh^3}{12} = \frac{(4.5)(0.0625)^3}{12}$$

$$= 9.155 \times 10^{-5}\ \text{in}^4$$

The distance from C's centroid to the composite centroid is

$$\bar{y} = \frac{0.0625}{2} + 0.00625 = 0.0375\ \text{in}$$

From the parallel axis theorem,

$$I_C = I_{ox} + A\bar{y}^2$$

$$= (9.155 \times 10^{-5}) + (0.28125)(0.0375)^2$$

$$= 4.871 \times 10^{-4}\ \text{in}^4$$

The total moment of inertia is

$$I_t = I_A + I_B + I_C$$

$$= 8.373 \times 10^{-4} + 1.221 \times 10^{-7} + 4.871 \times 10^{-4}$$

$$= \boxed{1.325 \times 10^{-3}\ \text{in}^4}$$

3. $\rho_{\text{cast iron}} = 0.256\ \text{lbm/in}^3$

$$m_{\text{rim}} = V\rho = \left(\frac{\pi}{4}\right)\left[(60)^2 - (48)^2\right](12)(0.256)$$

$$= 3126.9\ \text{lbm}$$

$$m_{\text{hub}} = \left(\frac{\pi}{4}\right)\left[(12)^2 - (6)^2\right](12)(0.256)$$

$$= 260.6\ \text{lbm}$$

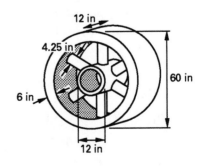

The arms are 18 in long.

$$m_{\text{arm}} = \left(\frac{\pi}{4}\right)(4.25)^2(18)(0.256)$$
$$= 65.4 \text{ lbm each}$$

For a hollow right circular cylinder,

$$J = \tfrac{1}{2}m(R_o^2 + R_i^2)$$

$$J_{\text{rim}} = \left(\tfrac{1}{2}\right)\left(\frac{3126.9}{32.2}\right)\left[\frac{\left(\frac{60}{2}\right)^2 + \left(\frac{48}{2}\right)^2}{144}\right]$$

$$= 497.7 \text{ slug-ft}^2$$

$$J_{\text{hub}} = \left(\tfrac{1}{2}\right)\left(\frac{260.6}{32.2}\right)\left[\frac{\left(\frac{12}{2}\right)^2 + \left(\frac{6}{2}\right)^2}{144}\right]$$

$$= 1.3 \text{ slug-ft}^2$$

The centroidal moment of inertia of a circular cylinder is

$$J_c = \left(\frac{1}{12}\right)m(3r^2 + L^2)$$

$$= \left(\frac{1}{12}\right)\left(\frac{65.4}{32.2}\right)\left[\frac{(3)\left(\frac{4.25}{2}\right)^2 + (18)^2}{144}\right]$$

$$= 0.4 \text{ slug-ft}^2 \text{ each}$$

The moment of inertia about the rotational axis (15 in away from the arm's centroid) is

$$J_{\text{arm}} = J_c + mr^2$$
$$= 0.4 + \left(\frac{65.4}{32.2}\right)\left(\frac{15}{12}\right)^2$$
$$= 3.57 \text{ slug-ft}^2 \text{ each}$$

For 6 arms,

$$_{\text{arms}} = (6)(3.57) = 21.4 \text{ slug-ft}^2$$
$$J_{\text{total}} = J_{\text{rim}} + J_{\text{hub}} + J_{\text{arms}}$$
$$= 497.7 + 1.3 + 21.4$$
$$= \boxed{520.4 \text{ slug-ft}^2}$$

Concentrates

1. For triangle EBD,

$$BE = \sqrt{(13.4)^2 - (12)^2}$$
$$= 5.96 \quad [\text{say } 6]$$
$$AE = 10 - 6 = 4$$

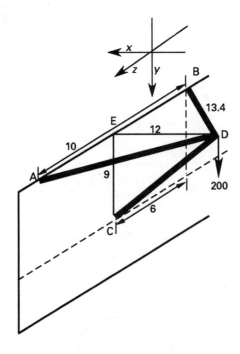

Using point D as the origin, the coordinates of the other points are

A: (12,0,4)
B: (12,0,−6)
C: (12,9,0)
D: (0,0,0)

The direction cosines of the applied load are

$$\cos\theta_x = 0$$
$$\cos\theta_y = 1$$
$$\cos\theta_z = 0$$

So,

$$F_x = 0$$
$$F_y = 200$$
$$F_z = 0$$

The lengths of the legs are

$$AD = \sqrt{(12-0)^2 + (0-0)^2 + (4-0)^2} = 12.65$$
$$BD = 13.4 \quad [\text{given}]$$
$$CD = \sqrt{(12)^2 + (9)^2 + (0)^2} = 15.00$$

The leg direction cosines are as follows.

For AD,

$$\cos\theta_{xA} = \frac{12}{12.65} = 0.949$$

$$\cos\theta_{yA} = \frac{0}{12.65} = 0$$

$$\cos\theta_{zA} = \frac{4}{12.65} = 0.316$$

For BD,

$$\cos\theta_{x\mathrm{B}} = \frac{12}{13.4} = 0.896$$

$$\cos\theta_{y\mathrm{B}} = \frac{0}{13.4} = 0$$

$$\cos\theta_{z\mathrm{B}} = \frac{-6}{13.4} = -0.448$$

For CD,

$$\cos\theta_{x\mathrm{C}} = \frac{12}{15.00} = 0.8$$

$$\cos\theta_{y\mathrm{C}} = \frac{9}{15.00} = 0.6$$

$$\cos\theta_{z\mathrm{C}} = \frac{0}{15.00} = 0$$

$$0.949F_\mathrm{A} + 0.896F_\mathrm{B} + 0.8F_\mathrm{C} + \quad 0 = 0$$
$$0.6F_\mathrm{C} + 200 = 0$$
$$-0.316F_\mathrm{A} - 0.448F_\mathrm{B} \qquad\quad + \quad 0 = 0$$

The simultaneous solution to these three equations is

$$\boxed{\begin{aligned} F_\mathrm{C} &= -333.3\ \mathrm{lbf} \quad [\text{compression}] \\ F_\mathrm{B} &= 118.9\ \mathrm{lbf} \quad [\text{tension}] \\ F_\mathrm{A} &= 168.6\ \mathrm{lbf} \quad [\text{tension}] \end{aligned}}$$

2. Taking counterclockwise moments as positive,

$$\sum M_c = D_y(6) - (8000)(6) + (1600)(16)$$
$$= 0$$
$$D_y = 3733\ \mathrm{kips}$$

Since DE is the only vertical member leaving point D,

$$\mathrm{DE} = \boxed{3733\ \mathrm{kips}\ [\text{compression}]}$$

3. Divide into three areas.

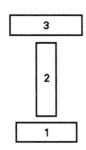

$$A_1 = (4)(1) = 4\ \mathrm{ft}^2$$
$$\bar{y}_1 = \tfrac{1}{2}\ \mathrm{ft}$$

$$A_2 = (2)(12) = 24\ \mathrm{ft}^2$$
$$\bar{y}_2 = 1 + 6 = 7\ \mathrm{ft}$$
$$A_3 = (6)\left(\tfrac{1}{2}\right) = 3\ \mathrm{ft}^2$$
$$\bar{y}_3 = 13.25\ \mathrm{ft}$$

$$\bar{y}_c = \frac{(4)(0.5) + (24)(7) + (3)(13.25)}{4 + 24 + 3} = 6.77\ \mathrm{ft}$$

$$I_{c1} = \frac{bh^3}{12} = \frac{(4)(1)^3}{12} = 0.333\ \mathrm{ft}^4$$
$$d_1 = 6.77 - 0.5 = 6.27\ \mathrm{ft}$$
$$I_{c2} = \frac{(2)(12)^3}{12} = 288\ \mathrm{ft}^4$$
$$d_2 = 0.23\ \mathrm{ft}$$
$$I_{c3} = \frac{(6)\left(\tfrac{1}{2}\right)^3}{12} = 0.0625\ \mathrm{ft}^4$$
$$d_3 = 6.48\ \mathrm{ft}$$

Using the parallel axis theorem,

$$I_{\text{total}} = 0.333 + (4)(6.27)^2 + 288 + (24)(0.23)^2$$
$$+ 0.0625 + (3)(6.48)^2$$
$$= \boxed{572.9\ \mathrm{ft}^4}$$

4. Move the origin to the apex of the tripod. The coordinates of the tripod bases become

$$\begin{aligned} \mathrm{A}: &\quad (5, -12, 0) \\ \mathrm{B}: &\quad (0, -8, -8) \\ \mathrm{C}: &\quad (-4, -7, 6) \end{aligned}$$

By inspection, $F_x = 1200$, $F_y = 0$, and $F_z = 0$.

$$L_\mathrm{A} = \sqrt{(5)^2 + (-12)^2 + (0)^2} = 13$$

$$\cos\theta_{x\mathrm{A}} = \frac{5}{13} = 0.385$$

$$\cos\theta_{y\mathrm{A}} = \frac{-12}{13} = -0.923$$

$$\cos\theta_{z\mathrm{A}} = \frac{0}{13} = 0$$

Similarly,

$$L_\mathrm{B} = \sqrt{(0)^2 + (-8)^2 + (-8)^2} = 11.31$$

$$\cos\theta_{x\mathrm{B}} = \frac{0}{11.31} = 0$$

$$\cos\theta_{y\mathrm{B}} = \frac{-8}{11.31} = -0.707$$

$$\cos\theta_{z\mathrm{B}} = \frac{-8}{11.31} = -0.707$$

$$L_\mathrm{C} = \sqrt{(-4)^2 + (-7)^2 + (6)^2} = 10.05$$

$$\cos \theta_{xC} = \frac{-4}{10.05} = -0.398$$

$$\cos \theta_{yC} = \frac{-7}{10.05} = -0.697$$

$$\cos \theta_{zC} = \frac{6}{10.05} = 0.597$$

$$
\begin{aligned}
0.385 F_A \qquad\quad - 0.398 F_C &= -1200 \\
- 0.923 F_A - 0.707 F_B - 0.697 F_C &= 0 \\
- 0.707 F_B + 0.597 F_C &= 0
\end{aligned}
$$

Solving these simulanteously yields

$$
\boxed{
\begin{aligned}
F_A &= -1793 \text{ lbf} \quad \text{[compression]} \\
F_B &= 1080 \text{ lbf} \quad \text{[tension]} \\
F_C &= 1279 \text{ lbf} \quad \text{[tension]}
\end{aligned}
}
$$

5. By symmetry, $A_y = L_y = 160$ kips.

For DE, cut as shown and sum the vertical forces.

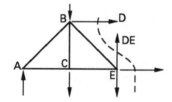

Taking upward forces as positive,

$$\sum F_y = 160 + \text{DE} - 60 - 60 - 4 = 0$$

$$\text{DE} = \boxed{-36 \text{ kips [compression]}}$$

For HJ, cut as shown and take moments about point I.

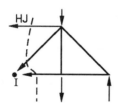

Taking counterclockwise moments as positive,

$$\sum M_{\text{I}} = (160)(60) - (60)(30) - (4)(30) + 20 \text{HJ} = 0$$

$$\text{HJ} = \boxed{-384 \text{ kips [compression]}}$$

6. In the x-y plane,

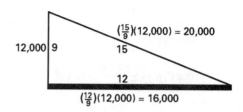

In the x-z plane,

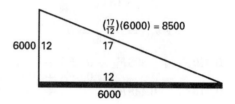

$$
\boxed{
\begin{aligned}
A_x &= 6000 + 16{,}000 = 22{,}000 \\
A_y &= 0 \\
A_z &= 0 \\
B_x &= 16{,}000 \\
B_y &= 12{,}000 \\
B_z &= 0 \\
C_x &= 6000 \\
C_y &= 0 \\
C_z &= 6000
\end{aligned}
}
$$

7.
$$S = c\left[\cosh\left(\frac{a}{c}\right) - 1\right]$$
$$w = 2 \text{ lbm/ft}$$
$$S = 10 \text{ ft}$$
$$a = 50 \text{ ft}$$
$$c = 126.6 \text{ ft} \quad \text{[trial and error]}$$

The midpoint (horizontal) tension is given by

$$H = wc = (2)(126.6) = \boxed{253.2 \text{ lbf}}$$

Since $y = c + S$ at the support, the endpoint (maximum) tension is given by

$$T = wy = w(c + S)$$
$$= (2)(126.6 + 10) = \boxed{273.2 \text{ lbf}}$$

8. From $T = wy$,

$$y = \frac{T}{w} = \frac{500}{2} = 250 \text{ ft at right support}$$

$$250 = c\left[\cosh\left(\frac{50}{c}\right)\right]$$

By trial and error,

$$c = 245 \text{ ft}$$

$$S = (245)\left[\cosh\left(\frac{50}{245}\right) - 1\right] = \boxed{5.12 \text{ ft}}$$

(Or, notice that $S = y - c = 5$.)

9. By inspection, $\bar{y} = 0$. To find $\bar{x}$, divide the object into three parts.

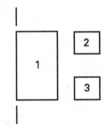

$$A_1 = (8)(4) = 32 \text{ ft}^2$$
$$\bar{x}_1 = 2 \text{ ft}$$
$$A_2 = A_3 = (2)(4) = 8 \text{ ft}^2$$
$$\bar{x}_2 = \bar{x}_3 = 6 \text{ ft}$$
$$\bar{x}_c = \frac{(32)(2) + (8)(6) + (8)(6)}{32 + 8 + 8} = \boxed{3.333 \text{ ft}}$$

10. For the parabola:

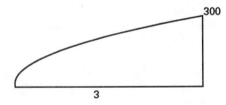

$$A = \frac{2bh}{3} = \frac{(2)(300)(3)}{3} = \boxed{600 \text{ lbf}}$$

The centroid is located

$$\frac{3h}{5} = \frac{(3)(3)}{5} = \boxed{1.8 \text{ ft from tip}}$$

Divide the remaining area into a triangle (T) and a rectangle (R).

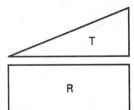

For the rectangle:

$$A_R = (8)(300) = \boxed{2400 \text{ lbf}}$$

$$\boxed{\text{The centroid is 4 ft from the right end.}}$$

For the triangle:

$$A_T = \left(\tfrac{1}{2}\right)(8)(700 - 300) = \boxed{1600 \text{ lbf}}$$

The centroid is located

$$\frac{8 \text{ ft}}{3} = \boxed{2.67 \text{ ft from the right end}}$$

Timed

1.

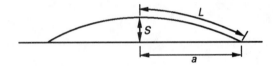

$$\alpha_{steel} = 6.5 \times 10^{-6} \text{ 1/°F}$$

$$\delta = \alpha L \Delta T$$
$$= \left(6.5 \times 10^{-6}\frac{1}{°F}\right)(5280 \text{ ft})(99.14°F - 70°F)$$
$$= 1.00008 \text{ ft} \approx 1 \text{ ft}$$
$$L = \frac{5280 + 1}{2} = 2640.5 \text{ ft}$$

For thermal expansion, assume that the distributed load is uniform along the length of the rail. This resembles the case of a cable under its own weight, the catenary. When distance S is small relative to distance a, the problem can be simplified using the parabolic solution. Given L (2604.5 ft) and a (2640 ft), find S by trial and error.

$$2640.5 \text{ ft} \approx (2640)\left[1 + \left(\frac{2}{3}\right)\left(\frac{S}{2640}\right)^2 - \left(\frac{2}{5}\right)\left(\frac{S}{2640}\right)^4\right]$$

$$\boxed{S \approx 44.5 \text{ ft}}$$

MATERIALS SCIENCE

Warmups

1. (a)

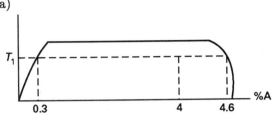

$$\alpha = \boxed{0.3\% \text{ A, } 99.7\% \text{ B}}$$

$$\beta = \boxed{4.6\% \text{ A, } 95.4\% \text{ B}}$$

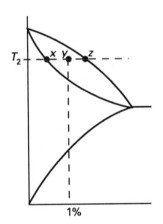

(b) Measure the line segment lengths.

$$xy = 9 \text{ mm}$$
$$xz = 21 \text{ mm}$$

$$\text{percent of liquid} = \frac{9}{21} = \boxed{0.43 \ (43\%)}$$

$$\text{percent of solid} = 100 - 43 = \boxed{57\%}$$

2. The change in diameter is

$$(0.3)(0.0200) = 0.006$$

$$\sigma_{\text{true}} = \frac{P}{A_o(1 - 0.006)^2}$$

$$= \frac{20,000 \text{ psi}}{(1 - 0.006)^2}$$

$$= \boxed{20,242 \text{ psi}}$$

$$\epsilon_{\text{true}} = \ln\left(\frac{L}{L_o}\right) = \ln(1 + 0.02)$$

$$= \boxed{0.0198 \text{ in/in}}$$

3.

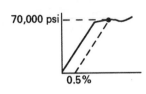

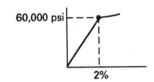

$$\text{slope} = \frac{60,000}{0.02} = 3 \times 10^6 \text{ psi}$$

The highest stress reached is approximately $\boxed{80 \text{ ksi}}$

The stress at the fracture is approximately $\boxed{70 \text{ ksi}}$

The strain at the fracture is approximately $\boxed{8\%}$

4.

$$G = \frac{E}{(2)(1 + \mu)} = \frac{3 \times 10^6}{(2)(1 + 0.3)}$$

$$= \boxed{1.15 \times 10^6 \text{ psi}}$$

5.

$$\text{ductility} = \frac{4 - 3.42}{4}$$

$$= \boxed{0.145 \ (14.5\%) \text{ reduction in area}}$$

6. Toughness is the area under the curve. Divide the area into (20 ksi)(1%) squares. There are about 25 squares.

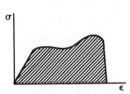

$$(25)\left(20,000 \ \frac{\text{lbf}}{\text{in}^2}\right)\left(0.01 \ \frac{\text{in}}{\text{in}}\right) = \boxed{5000 \text{ in-lbf/in}^3}$$

7. Plot the data and draw a straight line. Disregard the first and last data points.

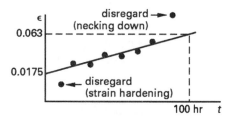

$$\text{slope} = \frac{0.063 - 0.0175}{100}$$

$$\boxed{= 0.000455 \ 1/\text{hr}}$$

8. The vinyl chloride mer is

$$
\begin{array}{ccc}
\text{H} & & \text{H} \\
| & & | \\
\text{C} & = & \text{C} \\
| & & | \\
\text{H} & & \text{Cl}
\end{array}
$$

The molecular weight is

$$(2)(12) + (3)(1) + (1)(35.5) = 62.5$$

With 20% efficiency, 5 molecules of HCl per PVC molecule are needed to supply the end $-$Cl atom. This is the same as 5 moles of HCl per mole of PVC.

$$\frac{(5)(6.022 \times 10^{23})}{7000} = \boxed{\begin{array}{l} 4.3 \times 10^{20} \\ \text{molecules HCl/gram PVC} \end{array}}$$

9. The molecular weights are

$$H_2O_2: (2)(1) + (2)(16) = 34 \text{ g/mol}$$
$$C_2H_4: (2)(12) + (4)(1) = 28 \text{ g/mol}$$

The number of H_2O_2 molecules in 10 ml is

$$\frac{(10 \text{ ml}) \left(1 \ \dfrac{\text{g}}{\text{ml}}\right) \left(\dfrac{0.2\%}{100}\right) \left(6.022 \times 10^{23} \ \dfrac{\text{molecules}}{\text{gmol}}\right)}{34 \ \dfrac{\text{g}}{\text{gmol}}}$$
$$= 3.54 \times 10^{20}$$

The number of ethylene molecules is

$$\frac{(12)(6.022 \times 10^{23})}{28} = 2.58 \times 10^{23}$$

Since it takes 1 H_2O_2 molecule (that is, 2 OH^- radicals) to stabilize a polyethylene molecule, there are 3.54×10^{20} polymers. The degree of polymerization is

$$\frac{2.58 \times 10^{23}}{3.54 \times 10^{20}} = \boxed{729 \text{ mers/polymer}}$$

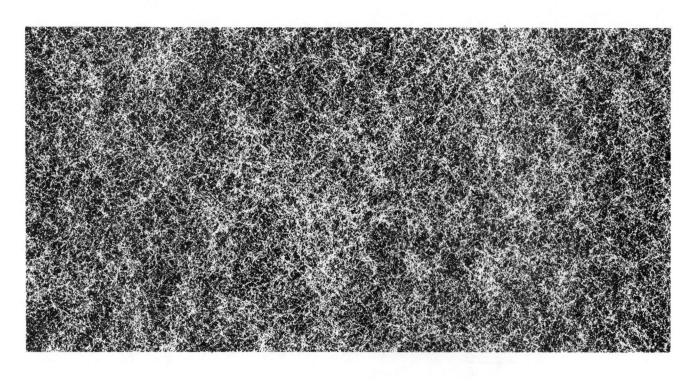

MECHANICS OF MATERIALS

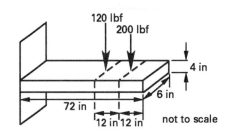

Warmups

1. First, find the reactions L and R.

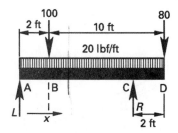

Taking the clockwise moment as positive,

$$\sum M_L: (100)(2) + (20)(12)(6)$$
$$+ (80)(12) - R(10) = 0$$
$$R = 260 \text{ lbf}$$

Taking upward forces as positive,

$$\sum F_y: L + 260 - 100 - 80 - (20)(12) = 0$$
$$L = 160 \text{ lbf}$$

Shear:

at A^+: $V = 160$
at B^-: $V = 160 - (2)(20) = 120$
at B^+: $V = 120 - 100 = 20$
at C^-: $V = 20 - (20)(8) = -140$
at C^+: $V = -140 + 260 = 120$
at D^-: $V = 120 - (20)(2) = 80$
at D^+: $V = 80 - 80 = 0$

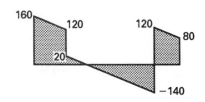

$$V_{\text{maximum}} = \boxed{160 \text{ lbf at the left end}}$$

Moment:

M_{maximum} occurs when $V = 0$. This occurs at

$$x = 2 + \left(\frac{20}{160}\right)(8) = 3.0 \text{ ft}$$

$$M_{\text{maximum}} = (160)(3) - (100)(1.0) - (20)(3)\left(\frac{3}{2}\right)$$

$$= \boxed{290 \text{ ft-lbf}}$$

2.

$$I = \frac{bh^3}{12} = \frac{(6)(4)^3}{12}$$
$$= 32 \text{ in}^4$$

For the 200-lbf load:

The deflection at the load is

$$y_1 = \frac{FL^3}{3EI} = \frac{(200)(60)^3}{(3)(1.5 \times 10^6)(32)} = 0.3000 \text{ in}$$

Differentiating y as a function of x gives the slope at any point.

$$y_1' = \left(\frac{F}{6EI}\right)(0 - 3L^2 + 3x^2)$$

At $x = 0$ and $L = 60$,

$$y_1' = \left[\frac{200}{(6)(1.5 \times 10^6)(32)}\right][(-3)(60)^2]$$
$$= -7.5 \times 10^{-3} \text{ in/in}$$

The added deflection at the tip is

$$y_2 = (7.5 \times 10^{-3})(12) = 0.0900 \text{ in}$$

For the 120-lbf load:

$$y_3 = \frac{(120)(48)^3}{(3)(1.5 \times 10^6)(32)} = 0.0922 \text{ in}$$
$$y_3' = \left[\frac{120}{(6)(1.5 \times 10^6)(32)}\right][(-3)(48)^2]$$
$$= -2.88 \times 10^{-3} \text{ in/in}$$
$$y_4 = (2.88 \times 10^{-3})(24) = 0.0691$$
$$y_{\text{tip}} = y_1 + y_2 + y_3 + y_4$$

$$= \boxed{0.5513 \text{ in}}$$

3.
$$E_{\text{steel}} = 29 \times 10^6 \text{ lbf/in}^2$$
$$\alpha = 6.5 \times 10^{-6} \text{ 1/°F}$$
$$\Delta L = \alpha L \Delta T$$
$$= (6.5 \times 10^{-6})(200)(12)(70)$$
$$= 1.092 \text{ in}$$

The constrained length is

$$1.092 - 0.5 = 0.592 \text{ in}$$

The strain is

$$\epsilon = \frac{\Delta L}{L} = \frac{0.592}{\left(200 + \dfrac{0.5}{12}\right)(12)}$$
$$= 2.466 \times 10^{-4}$$

The stress is

$$\sigma = \epsilon E = (2.466 \times 10^{-4})(2.9 \times 10^7)$$
$$= \boxed{7151 \text{ lbf/in}^2}$$

4. The design load is

$$(2.5)(75,000) = 187,500 \text{ lbf}$$

$C = 0.5$ for two fixed end.

$$L' = CL = (0.5)(50)(12) = 300 \text{ in}$$

The required moment of inertia is

$$I = \frac{FL^2}{\pi^2 E} = \frac{(187,500 \text{ lbf})(300 \text{ in})^2}{\pi^2 \left(2.9 \times 10^7 \dfrac{\text{lbf}}{\text{in}^2}\right)}$$
$$= \boxed{58.96 \text{ in}^4}$$

5. Assume soft steel.

$$E = 2.9 \times 10^7 \text{ lbf/in}^2$$
$$\Delta L = \frac{FL}{AE}$$
$$L_o = \frac{\Delta L A E}{F}$$
$$= \frac{(0.158 \text{ in})\left(\dfrac{\pi}{4}\right)(1 \text{ in})^2 \left(2.9 \times 10^7 \dfrac{\text{lbf}}{\text{in}^2}\right)}{15,000 \text{ lbf}}$$
$$= 239.913 \text{ in}$$
$$L_{\text{total}} = L_o + \Delta L$$
$$= 239.913 + 0.158$$
$$= \boxed{240.071 \text{ in}}$$

6. Assuming ASTM A36 steel, the yield strength is approximately 36,000 lbf/in^2.

$$FS = \frac{36,000}{8240} = \boxed{4.37}$$

Concentrates

1.

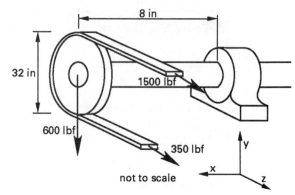

not to scale

For the shaft,

$$A = \frac{\pi}{4}d^2 = \left(\frac{\pi}{4}\right)(3)^2 = 7.07 \text{ in}^2$$
$$I = \frac{\pi}{4}r^4 = \left(\frac{\pi}{4}\right)(1.5)^4 = 3.98 \text{ in}^4$$
$$J = \frac{\pi}{2}r^4 = \left(\frac{\pi}{2}\right)(1.5)^4 = 7.95 \text{ in}^4$$

The moment at the bearing face due to the pulley weight is

$$M_y = (600 \text{ lbf})(8 \text{ in}) = 4800 \text{ in-lbf}$$

The stress is

$$\sigma_y = \frac{Mc}{I} = \frac{(4800)(1.5)}{3.98} = 1809 \text{ lbf/in}^2$$

The moment at the bearing face due to the belt tensions is

$$M_z = (1500 + 350)(8) = 14,800 \text{ in-lbf}$$

The stress is

$$\sigma_z = \frac{(14,800)(1.5)}{3.98} = 5578 \text{ lbf/in}^2$$

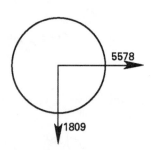

The maximum resultant bending stress is

$$\sigma_R = \pm\sqrt{(1809)^2 + (5578)^2} = \pm 5864 \text{ lbf/in}^2$$

The net torque is

$$T = (1500 - 350)(16) = 18{,}400 \text{ in-lbf}$$

The torsional shear stress is

$$\tau = \frac{Tc}{J} = \frac{(18{,}400)(1.5)}{7.95} = 3472 \text{ lbf/in}^2$$

Note: The direct shearing stress $\tau = V/A$ is zero on the shaft surface.

The maximum stress is

$$\left(\tfrac{1}{2}\right)5864 + \tfrac{1}{2}\sqrt{(5864)^2 + [(2)(3472)]^2}$$

$$= \boxed{7476 \text{ lbf/in}^2}$$

2. The radius of gyration is

$$k = \sqrt{\frac{I}{A}} = \sqrt{\frac{350 \text{ in}^4}{25.6 \text{ in}^2}}$$

$$= 3.70 \text{ in}$$

The slenderness ratio is

$$\frac{L}{k} = \frac{(25 \text{ ft})\left(12 \frac{\text{in}}{\text{ft}}\right)}{3.70 \text{ in}} = 81.08$$

$$\phi = \left(\tfrac{1}{2}\right)\left(\frac{L}{k}\right)\sqrt{\frac{F}{AE}}$$

$$= \left(\tfrac{1}{2}\right)(81.08)\sqrt{\frac{150{,}000 \text{ lbf}}{(25.6 \text{ in}^2)\left(2.9 \times 10^7 \frac{\text{lbf}}{\text{in}^2}\right)}}$$

$$= 0.5762 \text{ rad}$$

(b) The largest compressive stress is

$$\sigma_{max} = \left(\frac{F}{A}\right)\left[1 + \left(\frac{ec}{k^2}\right)\sec\phi\right]$$

$$= \left(\frac{150{,}000 \text{ lbf}}{25.6 \text{ in}^2}\right)$$

$$\times \left[1 + \left[\frac{(3.33 \text{ in})(7 \text{ in})}{(3.70 \text{ in})^2}\right]\sec(0.5762)\right]$$

$$= 17{,}757 \text{ lbf/in}^2$$

Note: 0.5762 is in radians not degrees.

The stress factor of safety is

$$(FS)_{stress} = \frac{S_y}{\sigma_{max}} = \frac{36{,}000 \frac{\text{lbf}}{\text{in}^2}}{17{,}757 \frac{\text{lbf}}{\text{in}^2}} = \boxed{2.03}$$

(a) Find the load that would cause the maximum stress to equal the yield stress.

$$S_y = \left(\frac{F}{A}\right)\left(1 + \left(\frac{ec}{k^2}\right)\sec\left[\left(\frac{Lk}{2}\right)\sqrt{\frac{F}{AE}}\right]\right)$$

$$36{,}000 \text{ lbf/in}^2 = \left(\frac{F_{max}}{25.6 \text{ in}^2}\right)\left(1 + \frac{(3.33 \text{ in})(7 \text{ in})}{(3.7 \text{ in})^2}\right.$$

$$\left.\times \sec\left[\frac{81.08}{2}\sqrt{\frac{F_{max}}{(25.6 \text{ in}^2)\left(2.9 \times 10^7 \frac{\text{lbf}}{\text{in}^2}\right)}}\right]\right)$$

$$36{,}000 \text{ lbf/in}^2 = \left(\frac{F_{max}}{25.6 \text{ in}^2}\right)\left(1 + 1.703\right.$$

$$\left.\times \sec\left[40.54\sqrt{\frac{F_{max}}{7.424 \times 10^8}}\right]\right)$$

Using a calculator's equation solver, or by trial and error,

$$F_{max} = 272{,}200 \text{ lbf}$$

The load factor of safety is

$$(FS)_{load} = \frac{F_{max}}{F} = \frac{272{,}200 \text{ lbf}}{150{,}000 \text{ lbf}}$$

$$= \boxed{1.81}$$

3. The largest face is 2 ft × 3 ft or 24 in × 36 in. Since the walls are fixed, flat plate equations for built-in edges may be used. For $a/b = 36/24 = 1.5$,

$$C_3 \approx \left(\tfrac{1}{2}\right)(0.436 + 0.487) = 0.462$$

$$\sigma_{max} \approx \frac{C_3 p b^2}{t^2} = \frac{(0.462)(2)(24)^2}{(0.25)^2}$$

$$= \boxed{8516 \text{ lbf/in}^2}$$

4.
$$r_1 = \frac{1.750}{2} = 0.875 \text{ in}$$

Ignore the end effects. (If this is not true, a combined stress analysis is required.) The maximum stress will occur at the inside. Using an allowable stress of 20,000 lbf/in^2,

$$\sigma_{c_i} = 20{,}000$$

$$= \frac{[r_o^2 + (0.875)^2](2000)}{(r_o + 0.875)(r_o - 0.875)}$$

$$(20{,}000)\left[r_o^2 - (0.875)^2\right] = \left[r_o^2 + (0.875)^2\right](2000)$$

$$r_o^2 = \left(\frac{11}{9}\right)(0.875)^2$$

$$\boxed{\begin{array}{l} r_o = 0.9673 \text{ in} \\ d_o = 1.935 \text{ in} \end{array}}$$

5. $$r_o = \frac{1.486}{2} = 0.743 \text{ in}$$

$$r_i = \frac{0.742}{2} = 0.371 \text{ in}$$

$$\sigma_{c_i} = \frac{(-2)(0.743)^2(400)}{(0.743 + 0.371)(0.743 - 0.371)}$$

$$= \boxed{-1066 \text{ lbf/in}^2 \text{ [compressive]}}$$

6.

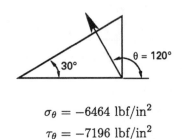

$$A = \left(\frac{\pi}{4}\right)(d^2) = \left(\frac{\pi}{4}\right)(1)^2 = 0.7854 \text{ in}^2$$

$$I = \left(\frac{\pi}{4}\right)(0.5)^4 = 0.04909 \text{ in}^4$$

$$J = \left(\frac{\pi}{2}\right)(0.5)^4 = 0.09817 \text{ in}^4$$

The moment at the chuck is *(bending force)*

$$(60 \text{ lbf})(8 \text{ in}) = 480 \text{ in-lbf}$$

$$\sigma = \frac{Mc}{I} = \frac{(480)(0.5)}{0.04909}$$

$$= 4889 \text{ lbf/in}^2$$

The applied torque is

$$T = (60 \text{ lbf})(12 \text{ in})$$

$$= 720 \text{ in-lbf}$$

The shear stress is

$$\tau = \frac{Tc}{J} = \frac{(720)(0.5)}{0.09817}$$

$$= 3667 \text{ lbf/in}^2$$

$$\tau \quad \mathcal{T}_{max} = \frac{1}{2}\sqrt{(4889)^2 + [(2)(3667)]^2} \quad \text{EQ. 14.47}$$

$$= 4407 \text{ lbf/in}^2$$

$$\sigma_{max} = \left(\frac{1}{2}\right)(4889) + 4407 \quad \text{EQ. 14.46}$$

$$= \boxed{6852 \text{ lbf/in}^2}$$

7.

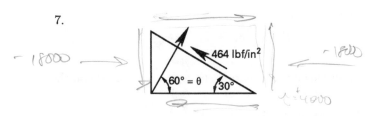

$$\sigma_1 = \frac{-18,000}{1.5} = -12,000 \text{ lbf/in}^2$$

(negative because compressive)

$$\sigma_2 = 0$$

For a 30° incline, $\theta = 60°$.

$$\sigma_\theta = \left(\frac{1}{2}\right)(-12,000) + \left(\frac{1}{2}\right)(-12,000)\cos 120°$$
$$\quad + (4000)\sin 120°$$
$$= 464 \text{ lbf/in}^2$$

$$\tau_\theta = -\left(\frac{1}{2}\right)(-12,000)\sin 120° + (4000)\cos 120°$$

$$= \boxed{3196 \text{ lbf/in}^2}$$

Alternative interpretation:

$$\sigma_\theta = -6464 \text{ lbf/in}^2$$
$$\tau_\theta = -7196 \text{ lbf/in}^2$$

8. The area in tension is

$$\left(\frac{\pi}{4}\right)\left[(16)^2 - (15.8)^2\right] = 5 \text{ in}^2$$

$$\sigma_N = \frac{F}{A} = \frac{40,000}{5}$$
$$= 8000 \text{ lbf/in}^2$$

$$J = \left(\frac{1}{32}\right)\pi\left[(16)^4 - (15.8)^4\right]$$
$$= 315.7 \text{ in}^4$$

$$\tau = \frac{Tc}{J} = \frac{(400,000)(8)}{315.7}$$
$$= 10,136 \text{ lbf/in}^2$$

$$\tau_{\max} = \tfrac{1}{2}\sqrt{(8000)^2 + [(2)(10{,}136)]^2}$$
$$= 10{,}897 \text{ psi}$$
$$\sigma_N = \left(\tfrac{1}{2}\right)(8000) \pm 10{,}897 \text{ lbf/in}^2$$
$$= \boxed{+\,14{,}897,\ -\,6897 \text{ lbf/in}^2}$$

Timed

1.

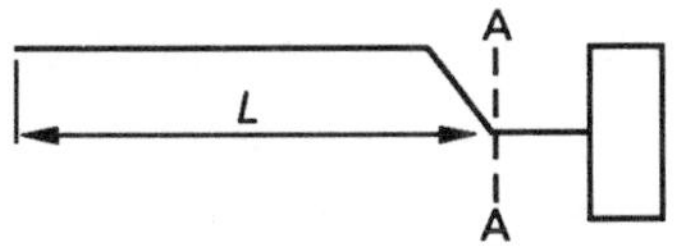

$$L = 14 + 3 = 17 \text{ in}$$
$$I = \left(\frac{\pi}{4}\right)(r)^4 = \left(\frac{\pi}{4}\right)\left[\frac{5}{(8)(2)}\right]^4 = 0.00749 \text{ in}^4$$
$$J = \left(\frac{\pi}{2}\right)(r)^4 = \left(\frac{\pi}{2}\right)\left[\frac{5}{(8)(2)}\right]^4 = 0.01498 \text{ in}^4$$

The moment at section A-A is

$$M = FL = (50)(17) = 850 \text{ in-lbf}$$

The bending stress is

$$\sigma = \frac{Mc}{I} = \frac{\dfrac{(850)\left(\frac{5}{8}\right)}{2}}{0.00749}$$
$$= \boxed{35{,}464 \text{ lbf/in}^2}$$

The torque at section A-A due to the eccentricity is

$$T = (50)(3) = 150 \text{ in-lbf}$$

The shear stress is

$$\tau = \frac{Tc}{J} = \frac{\dfrac{(150)\left(\frac{5}{8}\right)}{2}}{0.01498}$$
$$= 3129 \text{ lbf/in}^2$$

The maximum shear stress is

$$\tau_{\max} = \tfrac{1}{2}\sqrt{(35{,}464)^2 + [(2)(3129)]^2}$$
$$= 18{,}006 \text{ lbf/in}^2 \quad \left[\text{F/A direct shear is ignored}\right]$$
$$\sigma_1, \sigma_2 = \left(\tfrac{1}{2}\right)(35{,}464) \pm 18{,}006$$
$$= \boxed{35{,}738,\ -274 \text{ lbf/in}^2}$$

2. For the jacket,

$$r_o = \frac{12}{2} = 6 \text{ in}$$
$$r_i = \frac{7.75}{2} = 3.875 \text{ in}$$

For the tube,

$$r_o = \frac{7.75}{2} = 3.875 \text{ in}$$
$$r_i = \frac{4.7}{2} = 2.35 \text{ in}$$

For the jacket, this is no different than exposure to an internal pressure, p. σ_{ci} is the highest stress, so,

$$\sigma_{ci} = 18{,}000 = \frac{\left[(6)^2 + (3.875)^2\right]p}{(6 + 3.875)(6 - 3.875)}$$

The equivalent pressure and radial stress are

$$p = 7404 \text{ lbf/in}^2$$
$$\sigma_{r_i} = -7404 \text{ lbf/in}^2$$

Poisson ratio

$$\Delta D = \left(\frac{D}{E}\right)\left[\sigma_c - \mu(\sigma_r + \sigma_L)\right]$$
$$= \left(\frac{7.75}{29.6 \times 10^6}\right)[18{,}000 - (0.3)(-7404)]$$
$$= 0.00529 \text{ in}$$

Of course, the tube is also subjected to the pressure. So,

$$\sigma_{ro} = -7404 \text{ lbf/in}^2$$
$$\sigma_{co} = \left(\frac{-\left[(3.875)^2 + (2.35)^2\right]}{(3.875)^2 - (2.35)^2}\right)(7404)$$
$$= -16{,}018 \text{ lbf/in}^2$$

(Notice that $(r_o + r_i)t = (r_o + r_i)(r_o - r_i) = r_o^2 - r_i^2$.)

$$\Delta D = \left(\frac{7.75}{29.6 \times 10^6}\right)[-16{,}018 - (0.3)(-7404)]$$
$$= -0.00361 \text{ in}$$

The total interference is

$$0.00529 + 0.00361 = 0.0089 \text{ in}$$

(b) Maximum stress occurs at the inner jacket face.

$$\sigma_{ci} = 18{,}000 \text{ lbf/in}^2$$

(c) Minimum stress occurs at the outer tube face.

$$\sigma_{co} = \boxed{-16{,}018 \text{ lbf/in}^2}$$

3.

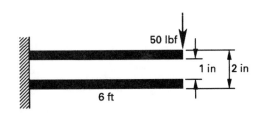

(a) Yes, the suggestion has merit. Anything in the center void will have a higher modulus of elasticity (stiffness) than air.

(b) Calculate I as the difference between two circles:

$$I_{\text{brass}} = \left(\frac{\pi}{4}\right)\left[\left(\frac{2}{2}\right)^4 - \left(\frac{1}{2}\right)^4\right] = 0.736 \text{ in}^4$$

$$E_{\text{brass}} = 15 \times 10^6 \text{ lbf/in}^2 = 1.5 \times 10^7 \text{ lbf/in}^2$$

$$EI_{\text{brass}} = (1.5 \times 10^7)(0.736) = 1.104 \times 10^7$$

$$I_{\text{steel}} = \left(\frac{\pi}{4}\right)\left(\frac{1}{2}\right)^4 = 0.0491 \text{ in}^4$$

Assume soft steel. So,

$$E_{\text{steel}} = 2.9 \times 10^7 \text{ lbf/in}^2$$

$$EI_{\text{steel}} = (2.9 \times 10^7)(0.0491) = 0.1424 \times 10^7$$

$$y_{\text{tip}} = \frac{FL^3}{3EI} \quad \left[\begin{array}{c}\text{inversely propor-}\\ \text{tional to EI}\end{array}\right]$$

The percent change is

$$\% = \frac{y_{\text{old}} - y_{\text{new}}}{y_{\text{old}}}$$

$$= \frac{\dfrac{1}{1.104} - \dfrac{1}{0.1424 + 1.104}}{\dfrac{1}{1.104}}$$

$$= \boxed{0.114 \ (11.4\%)}$$

4. S_y for 6061 T4 aluminum is approximately 19,000 lbf/in^2.

The following stresses are present in the spool.

- long stress due to pressure on end disks
- compressive hoop stress
- bending stress at disk and tube junction (end moment)
- circumferential (tangential) stress

Use the following assumptions.

- Bending stress is ignored.
- Circumferential stress is negligible due to thin-wall construction.
- Factor of safety calculation does not depend on obscure collapsing pressure theories, but rather on basic concepts.

The calculations are as follows.

$$\text{end disk area} = \left(\frac{\pi}{4}\right)\left[(2)^2 - (1)^2\right]$$

$$= 2.356 \text{ in}^2$$

$$\text{long force} = pA = (500)(2.356)$$

$$= 1178 \text{ lbf}$$

$$\begin{array}{c}\text{annulus area absorbing}\\ \text{the long force}\end{array} = \left(\frac{\pi}{4}\right)\left[(1)^2 - (1 - 0.10)^2\right]$$

$$= 0.149 \text{ in}^2$$

$$\sigma_{\text{long}} = \frac{1178}{0.149}$$

$$= 7906 \text{ lbf/in}^2 \quad \text{[tensile]}$$

$$\sigma_{\text{hoop}} = \frac{pr}{t} = \frac{(500)\left(\frac{1}{2}\right)}{0.050}$$

$$= \boxed{-5000 \text{ lbf/in}^2 \quad \text{[compressive]}}$$

These are the principal stresses. From the distortion energy theory, the von Mises stress is

$$\sigma' = \sqrt{\sigma_1^2 - \sigma_1\sigma_2 + \sigma_2^2}$$

$$= \sqrt{(7906)^2 - (7906)(-5000) + (-5000)^2}$$

$$= 11{,}271 \text{ lbf/in}^2$$

$$\text{FS} = \frac{19{,}000}{11{,}271} = 1.69$$

$$\boxed{\begin{array}{c}\text{The answer will depend}\\ \text{on the value of } S_y \text{ used.}\end{array}}$$

5.

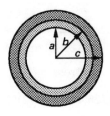

Put all lengths in terms of radius in meters.

$$a = \left(\frac{120 - 2 - 2 - 3 - 3}{2}\right) \times 10^{-3}$$

$$= 5.5 \times 10^{-2} \text{ m}$$

$$b = \left(\frac{120 - 2 - 2}{2}\right) \times 10^{-3}$$

$$= 5.8 \times 10^{-2} \text{ m}$$

$$c = \left(\frac{120}{2}\right) \times 10^{-3} = 6.0 \times 10^{-2} \text{ m}$$

$$I = 0.3 \times 10^{-3} = 3 \times 10^{-4} \text{ m}$$

interference

In general, the interference, I, is

$$I = |\Delta D_{\text{inner}}| + |\Delta D_{\text{outer}}|$$

ΔD is calculated the same for both. There is no long stress, so

$$\Delta D = \left(\frac{D}{E}\right)(\sigma_c - \mu\sigma_r)$$

For the inner cylinder,

- external pressure
- $r_i = a = 5.5 \times 10^{-2}$ m
- $r_o = b = 5.8 \times 10^{-2}$ m
- $t = b - a = 0.3 \times 10^{-2}$ m
- $D = 2b = 0.116$ m

$$\sigma_{co} = \frac{-(r_o^2 + r_i^2)p}{(r_o + r_i)t}$$

$$= \frac{-\left[(5.8)^2 + (5.5)^2\right](10^{-2})^2 p}{(5.8 + 5.5)(0.3)(10^{-2})^2}$$

$$= -18.85p$$

$$\sigma_{ro} = -p$$

$$\Delta D = \left(\frac{0.116}{207 \times 10^9}\right)[-18.85p - (0.3)(-p)]$$

$$= \boxed{-(1.04 \times 10^{-11})p}$$

For the outer cylinder,

- internal pressure
- $r_i = b = 5.8 \times 10^{-2}$ m
- $r_o = c = 6.0 \times 10^{-2}$ m
- $t = 0.2 \times 10^{-2}$ m
- $D = 2c = 0.120$ m

$$\sigma_{ri} = -p$$

$$\sigma_{ci} = \frac{(r_o^2 + r_i^2)p}{(r_o + r_i)t}$$

$$= \frac{\left[(6)^2 + (5.8)^2\right]p}{(6.0 + 5.8)(0.2)}$$

$$= 29.51p$$

$$\Delta D = \left(\frac{0.116}{207 \times 10^9}\right)[29.51p - (0.3)(-p)]$$

$$= +(1.67 \times 10^{-11})p$$

Since I is known to be 3×10^{-4} m,

$$3 \times 10^{-4} = p\left[(1.04 \times 10^{-11}) + (1.67 \times 10^{-11})\right]$$

$$p = 1.11 \times 10^7 \text{ N/m}^2 \text{ (pascals)}$$

Now that p is known, the circumferential stresses can be found.

For the inner cylinder,

$$\boxed{\sigma_{co} = -18.85p = -2.09 \times 10^8 \text{ N/m}^2 \quad \text{[compressive]}}$$

For the outer cylinder,

$$\boxed{\sigma_{ci} = 29.51p = 3.28 \times 10^8 \text{ N/m}^2 \quad \text{[tensile]}}$$

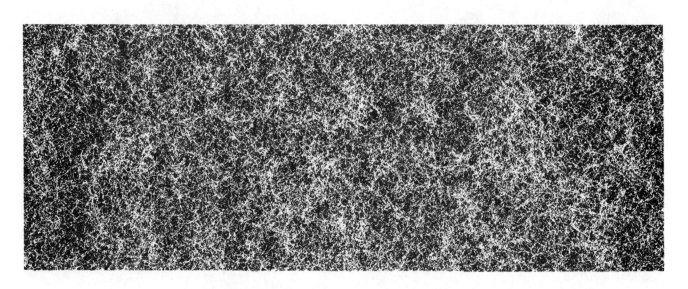

PROFESSIONAL PUBLICATIONS, INC. ● Belmont, CA

MACHINE DESIGN

Warmups

1.

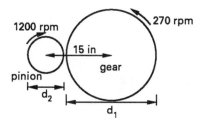

First, find the pitch diameters. Let gear 2 be the pinion and gear 1 be the gear. Then,

$$\frac{d_1}{2} + \frac{d_2}{2} = 15 \text{ in}$$

$$\frac{d_1}{d_2} = \frac{1200}{270}$$

Solving these simultaneously,

$$d_1 = 24.49 \text{ in}$$

$$d_2 = 5.51 \text{ in}$$

The pitch line velocity is

$$v = \frac{\pi(\text{rpm})d}{12} = \frac{\pi(270)(24.49)}{12}$$

$$= 1731.1 \text{ ft/min}$$

This velocity is the same for both gears.

Assume for the pinion that $w = 6$ and $Y = 0.2$.

Since no hardness data was given, the endurance limit is chosen as the maximum stress. Assume for the 1045 steel that $S_{\text{endurance}} = 90$ ksi.

Use a factor of safety of 3. The allowable stress is

$$\sigma_{\text{allowable}} = \frac{90}{3} = 30 \text{ ksi}$$

The Barth speed factor is

$$k_d = \frac{600}{600 + 1731.1} = 0.257$$

Neglecting the small amount of radial stress,

$$\sigma = \frac{(\text{hp})(33,000)P}{v k_d w Y}$$

$$P = \frac{(0.257)(30,000)(6)(0.2)(1731.1)}{(550)(33,000)}$$

$$= 0.88$$

This is unrealistic, if not impossible ($P = 2$ is a practical minimum). Continue procedurally, anyway.

Choose the number of teeth per inch as 1. The number of teeth on the pinion and gear are

$$N_{\text{pinion}} = Pd_p = (1)(5.51) = 5.51 \quad [\text{say 6}]$$

$$N_{\text{gear}} = (1)(24.49) = 24.49 \quad [\text{say 25}]$$

Now that the approximate numbers of teeth are known, the actual form factors can be determined.

$$Y_{\text{pinion}} \approx 0.15$$

$$Y_{\text{gear}} \approx 0.303$$

The new values can be used to find w.

$$w = \frac{(\text{hp})(33,000)(P)}{v \sigma Y k_d}$$

$$w_{\text{pinion}} = \frac{(550)(33,000)(1)}{(1731.1)(30,000)(0.15)(0.257)}$$

$$= 9.07 \text{ in}$$

Assume $\sigma_{\text{max}} = 50,000$ psi for the cast steel gear.

$$w_{\text{gear}} = \frac{(550)(33,000)(1)}{(1731.1)\left(\dfrac{50,000}{3}\right)(0.303)(0.257)}$$

$$= 8.08 \text{ in}$$

This disregards stress concentration factors. Both gears would be 9 in or larger. This meets the general rule of widths being 3–5 times π/P.

2. The maximum speed ratio is

$$\frac{96}{12} = 8$$

With 3 stages, $(8)^3 = 512$. This is too small. With 4 stages, $(8)^4 = 4096$.

Four sets of gears are needed. Try to keep all the gear sets as close as possible.

$$\sqrt[4]{600} \approx 5$$

$$\left(\frac{60}{12}\right)\left(\frac{60}{12}\right)\left(\frac{60}{12}\right)\left(\frac{72}{15}\right) = 600$$

Other combinations are possible.

3. This subject is not covered in the *Mechanical Engineering Reference Manual*.

For a B v-belt, the correction to the inside circumference is 1.8 inches. The pitch length is

$$L_p = L_{\text{inside}} + \text{correction} = 90 \text{ in} + 1.8 \text{ in}$$
$$= 91.8 \text{ in}$$

The ratio of speeds determines the sheave sizes.

$$d_{\text{equipment}} = d_{\text{motor}} \left(\frac{n_{\text{motor}}}{n_{\text{equipment}}} \right)$$
$$= (10 \text{ in}) \left(\frac{1750 \text{ rpm}}{800 \text{ rpm}} \right)$$
$$= \boxed{21.875 \text{ in}}$$

The closest standard sheave size is 21.8 in. This is more than the minimum diameter (5.4 in) for a B v-belt.

4.

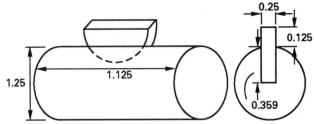

The above dimensions are taken from *Machinery's Handbook*.

For 1030 steel,

$$S_{yt} = 73,900 \text{ psi}$$
$$S_{ys} = (0.577)(73,900) = 42,640 \text{ psi}$$
$$S_{yc} \approx S_{yt} = 73,900 \text{ psi}$$

There are two possible approaches to this problem. The first method would have you calculate the stresses based on the actual areas in shear and bearing. This is almost never done. Rather, approximations are used. If the factor of safety is too small, use more than one key.

The usual aproach is to assume the entire length of the key carries the shear load. The area in shear is

$$A = Lw = (1.125)(0.25) = 0.2813 \text{ in}^2$$

Assuming the torque is carried at the shaft surface, the maximum torque would be

$$T = Fr = S_{ys}Ar$$
$$= (42,640)(0.2813) \left(\frac{1.25}{2} \right)$$
$$= 7497 \text{ in-lbf}$$

The factor of safety in shear is

$$(\text{FS})_s = \frac{7497}{4200} = \boxed{1.79}$$

The smallest area in bearing occurs in the hub and is

$$A = Ld = (1.125)(0.125) = 0.1406 \text{ in}^2$$

The actual applied force is

$$F = \frac{T}{4} = \frac{4200 \text{ in-lbf}}{\dfrac{1.25 \text{ in}}{2}} = 6720 \text{ lbf}$$

The bearing stress is

$$\sigma = \frac{F}{A} = \frac{6720}{0.1406} = 47,795 \text{ psi}$$

Using the compressive yield strength as the maximum bearing stress, the factor of safety in bearing is

$$(\text{FS})_b = \frac{73,900}{47,795} = \boxed{1.55}$$

5.

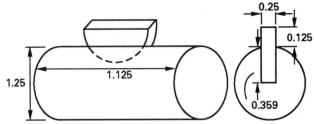

The area of the pin in shear is

$$(2 \text{ surfaces}) \left(\frac{\pi}{4} \right) d^2 = (1.571d^2) \text{ in}^2$$

The total shear is

$$V = \frac{T}{r} = \frac{400 \text{ in-lbf}}{\dfrac{1.125 \text{ in}}{2}} = 711.1 \text{ lbf}$$

For 1030 steel,

$$S_{yt} = 73,900 \text{ psi}$$
$$S_{ys} = (0.577)(73,900) = 42,640 \text{ psi}$$

Use a factor of safety of 2.5.

$$\tau_{\text{max}} = \frac{42,640}{2.5} = 17,060 \text{ psi}$$

The maximum shear stress experienced by a round bar in shear is

$$\frac{4V}{3A} = \frac{(4)(711.1 \text{ lbf})}{(3)(2)\left(\frac{\pi}{4}\right)d^2} = 17,060 \text{ lbf/in}^2$$

$$d = \boxed{0.188 \text{ in}}$$

6. *Annealing*: Heat to above the recrystalization temperature (600°F) and air cool to room temperature. The *ALCOA Aluminum Handbook* recommends 775°F held for 2–3 hours, cooling at 50°F/hr down to 500°F. The subsequent cooling rate is not important.

Precipitation hardening (based on the *ALCOA Aluminum Handbook*):

1. Solution treatment at 950°F for about 4 hours followed by rapid quench in cold water.

2. Reheat to 320°F for 12–16 hours, followed by cooling at a rate that is not unduly slow.

7.

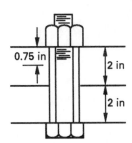

Assume $E = 2.9 \times 10^7$ psi. Neglect plate compression and nut and head deformation.

The body of the bolt has a diameter and area of

$$d = 0.75 \text{ in}$$

$$A = \left(\frac{\pi}{4}\right)(0.75)^2 = 0.4418 \text{ in}^2$$

The stress area is

$$A = 0.3724 \text{ in}^2$$

The elongation in the unthreaded part of the bolt is

$$\delta = \frac{FL}{AE} = \sigma\frac{L}{E}$$

$$= \frac{(40,000 \text{ psi})(3.25 \text{ in})}{2.9 \times 10^7 \text{ psi}}$$

$$= 0.00448 \text{ in}$$

The stress in the threaded part of the bolt is

$$(40,000)\left(\frac{0.4418}{0.3724}\right) = 47,454 \text{ psi}$$

No information about the nut is given, but half of the threaded section in the nut can be assumed to contribute to elongation. Arbitrarily take the length of the threaded section as three threads.

The elongation in the threaded part, including three threads in the nut, is

$$\delta = \sigma\frac{L}{E} = \frac{(47,454)\left[0.75 + (3)\left(\frac{1}{16}\right)\right]}{2.9 \times 10^7}$$

$$= 0.00153 \text{ in}$$

The total elongation is

$$\delta_t = 0.00448 + 0.00153 = \boxed{0.00601 \text{ in}}$$

8. The bolt must be loaded in simple tension. It cannot be used in a clamping configuration, since a preload would be required to prevent joint separation.

$$F_{\text{ave}} = \left(\tfrac{1}{2}\right)(1000 + 8000) = 4500 \text{ lbf}$$
$$F_{\text{alt}} = \left(\tfrac{1}{2}\right)(8000 - 1000) = 3500 \text{ lbf}$$

The threaded section has a minimum area of

$$A_{\text{min}} = 0.2256 \text{ in}^2$$

The stress concentration factor for rolled threads is 2.2 (Shigley and Mischke, *Mechanical Engineering Design*, 5th Edition, p. 351).

$$\sigma_{\text{ave}} = \frac{4500}{0.2256} = 19,947 \text{ psi}$$

$$\sigma_{\text{alt}} = \frac{(3500)(2.2)}{0.2256} = 34,131 \text{ psi}$$

Since S_{ut} is not known, the Soderberg diagram must be used.

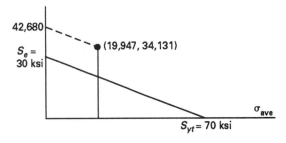

Failure will occur because the point is above the line. The factor of safety is

$$\text{FS} \approx \frac{30,000}{42,680} = \boxed{0.703}$$

9. The deflection is

$$y = \frac{FL^3}{3EI}$$

$$I = \frac{bh^3}{12} = \frac{6h^3}{12} = 0.5h^3$$

$$L = (2)(12) = 24 \text{ in}$$

$$2 \text{ in} = \frac{(800)(24)^3}{(3)(2.9 \times 10^7)(0.5)h^3}$$

$$h = \boxed{0.503 \text{ in}}$$

The width-thickness ratio is

$$\frac{w}{t} = \frac{6.0 \text{ in}}{0.503 \text{ in}} \approx 12$$

Since $w/t > 10$, this is a wide beam, and the deflection will be less than predicted by the regular beam equation. The reduction is $1 - \mu^2$, where Poisson's ratio is $\mu = 0.30$ for steel.

The required thickness is reduced.

$$\text{thickness} = (0.503 \text{ in})\left[1 - (0.3)^2\right]$$

$$= \boxed{0.458 \text{ in}}$$

The approximate bending stress is

$$\sigma = \frac{Mc}{I}$$

$$= \frac{(800 \text{ lbf})(2 \text{ ft})\left(12 \frac{\text{in}}{\text{ft}}\right)\left(\frac{0.458 \text{ in}}{2}\right)}{\frac{(6 \text{ in})(0.458)^3}{12}}$$

$$= 91{,}531 \text{ psi}$$

10. The change in kinetic energy is

$$\Delta K.E. = \frac{1}{2}J(\omega_1^2 - \omega_2^2)$$

$$= 2J\left(\frac{\pi}{60}\right)^2(n_1^2 - n_2^2) \quad [n \text{ in rpm}]$$

$$J = \frac{\Delta K.E.}{(2)\left(\frac{\pi}{60}\right)^2(n_1^2 - n_2^2)}$$

$$= \frac{1500 \text{ ft-lbf}}{(2)\left(\frac{\pi}{60}\right)^2\left[(200)^2 - (175)^2\right]}$$

$$= 29.18 \text{ ft-lbf-sec}^2 \quad [\text{slug-ft}^2]$$

Assuming all the mass is concentrated at the mean radius $r_m = 15$ in,

$$J = mr_m^2 = \frac{wtL_m\rho r_m^2}{g_c}$$

$$L_m = 2\pi r_m$$

The thickness is

$$t = \frac{g_c J}{2\pi w \rho r_m^3}$$

Including the 10% factor,

$$t = \frac{(32.2)(29.18)}{(1.10)\left(\frac{12}{12}\right)(2\pi)\left(\frac{15}{12}\right)^3(0.26)(12)^3}$$

$$= 0.155 \text{ ft} = \boxed{1.86 \text{ in}}$$

11. Assume a coefficient of friction of $\mu = 0.12$.

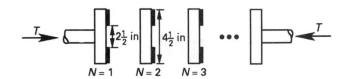

With N friction planes, the contact surface area is

$$A = N\left(\frac{\pi}{4}\right)\left[(4.5 \text{ in})^2 - (2.5 \text{ in})^2\right] = 11N \text{ in}^2$$

With a contact pressure of $p = 100$ psi, the normal force is

$$F_N = Ap = \left(11N \text{ in}^2\right)\left(100 \frac{\text{lbf}}{\text{in}^2}\right) = 1100N \text{ lbf}$$

The frictional force is

$$F_f = \mu F_N = (0.12)(1100N) \text{ lbf} = 132N \text{ lbf}$$

Assume the frictional force is applied at the mean radius.

$$r_m = \left(\frac{1}{2}\right)\left(\frac{2.5 + 4.5}{2}\right) = 1.75 \text{ in}$$

The resisting torque is

$$T_R + r_m F_f = (1.75 \text{ in})(132N \text{ lbf}) = 231N \text{ in-lbf}$$

Setting the slipping torque equal to the resisting torque,

$$(3)(300 \text{ in-lbf}) = 231N \text{ in-lbf}$$

$$N = \boxed{3.9}$$

Use 4 contact surfaces—3 plates and 2 discs.

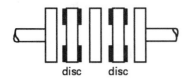

disc   disc

Concentrates

1. Design for static loading. Choose a spring index of 9.

$$d = \frac{8FC^3 N_a}{G\delta}$$

$$= \frac{(8)(50)(9)^3(12)}{(1.2 \times 10^7)(0.5)}$$

$$= 0.583 \text{ in}$$

$$D = 9d = (9)(0.583)$$

$$= \boxed{5.249 \text{ in}} \quad \begin{bmatrix} \text{a nonstandard} \\ \text{wire diameter} \end{bmatrix}$$

2. Assume ASTM A230 wire. Assume $d = 0.15$ in wire.

$$S_{ut} = 205,000 \text{ psi}$$

With a safety factor of 1, the allowable working stress is

$$\tau = \frac{(0.3)(S_{ut})}{\text{FS}} = \frac{(0.3)(205,000)}{1.5}$$

$$= 41,000$$

The Wahl factor is

$$W = \frac{(4)(10) - 1}{(4)(10) - 4} + \frac{0.615}{10} = 1.145$$

$$d = \sqrt{\frac{8CFW}{\pi\tau}} = \sqrt{\frac{(8)(10)(30)(1.145)}{\pi(41,000)}}$$

$$= 0.146 \text{ in}$$

Use W & M wire #9 with $d = 0.1483$ in.

$$D = 10d = (10)(0.1483) = 1.483 \text{ in}$$

The spring constant is

$$k = \frac{\Delta F}{\Delta \delta} = \frac{30 - 20}{0.3}$$

$$= 33.33 \text{ lbf/in}$$

$$N_a = \frac{Gd}{8kC^3} = \frac{(11.5 \times 10^6)(0.1483)}{(8)(33.33)(10)^3}$$

$$= 6.4 \text{ coils}$$

Specify a squared and ground spring. Use 8.4 coils with 6.4 active.

The solid height is $(6.4)(0.1483) = 0.949$ in.

At solid, the maximum stress should not exceed S_{us}.

$$S_{ut} = 205,000 \text{ psi}$$

For unpeened wire,

$$\tau_{\text{max,solid}} \approx 0.35 S_{ut} = (0.30)(205,000 \text{ psi})$$

$$= 61,500 \text{ psi}$$

The force at solid height is

$$F_{\text{solid}} = \frac{\tau_{\text{solid}} \pi d^3}{8DW}$$

$$= \frac{(61,500)\pi(0.1483)^3}{(8)(1.483)(1.145)}$$

$$= 46.39 \text{ lbf}$$

The deflection at solid is

$$\delta_{\text{solid}} = \frac{F_{\text{solid}}}{k} = \frac{46.39 \text{ lbf}}{33.33 \dfrac{\text{lbf}}{\text{in}}}$$

$$= 1.39 \text{ in}$$

The minimum free height is

$$0.949 \text{ in} + 1.39 \text{ in} = \boxed{2.34 \text{ in}}$$

3. The potential energy absorbed is

$$E_p = mg\Delta h$$

$$= \frac{(700)(32.2)(10 + 46)}{32.2}$$

$$= 39,200 \text{ in-lbf}$$

The work done by the spring is

$$W = \tfrac{1}{2}kx^2$$

$$39,200 = \tfrac{1}{2}k(10)^2$$

$$k = 784 \text{ lbf/in}$$

The equivalent force is

$$F = kx = (784)(10)$$

$$= 7840 \text{ lbf}$$

$$W = \frac{(4)(7) - 1}{(4)(7) - 4} + \frac{0.615}{7}$$

$$= 1.213$$

$$d = \sqrt{\frac{8CFW}{\pi\tau}}$$

$$= \sqrt{\frac{(8)(7)(7840)(1.213)}{\pi(50,000)}}$$

$$= 1.841 \text{ in}$$

$$D = 7d = (7)(1.841)$$

$$= 12.89 \text{ in}$$

$$N_a = \frac{Gd}{8kC^3} = \frac{(1.2 \times 10^7)(1.841)}{(8)(784)(7)^3}$$

$$= \boxed{10.27 \quad [\text{say } 10.3]}$$

4.

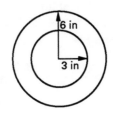

For the steel,

$$E_s = 2.9 \times 10^7 \text{ psi}$$

$$\mu_s = 0.3$$

For the cast iron,

$$E_c = 1.45 \times 10^7 \text{ psi}$$

$$\mu_c = 0.27$$

$$S_{ut} = 30,000 \text{ psi}$$

The only easy method of evaluating interference requires the stress-strain curve to be linear. For cast iron, non-linearity begins at approximately $(1/6)S_{ut}$. This sets the maximum stress for the cast iron, which is much less than the limiting steel stress.

$$\sigma < \frac{30,000}{6} = 5000 \text{ psi}$$

For the cast iron hub under internal pressure,

$$\sigma_{max} = \sigma_{ci} = 5000 \text{ psi}$$

$$\sigma_{ci} = 5000 = \frac{[(6)^2 + (3)^2]p}{[(6)^2 - (3)^2]}$$

$$p = 3000 \text{ psi}$$

$$\sigma_{ri} = -p$$

$$= -3000 \quad [\text{at the inner surface}]$$

The radial interference in the cast iron is

$$\Delta r = \left(\frac{r}{E}\right)[\sigma_c - \mu(\sigma_r + \sigma_L)]$$

$$= \left(\frac{3}{1.45 \times 10^7}\right)[5000 - (0.27)(-3000)]$$

$$= 1.20 \times 10^{-3} \text{ in}$$

For a cylinder under external pressure, σ_{ro} is maximum.

$$\sigma_{co} = (3000)\left(\frac{-[(3)^2 + (0)^2]}{(3)^2 - (0)^2}\right)$$

$$= -3000 \text{ psi}$$

$$\sigma_{ro} = -p = -3000$$

The radial interference in the steel is

$$\Delta r = \left(\frac{3}{2.9 \times 10^7}\right)[-3000 - (0.3)(-3000)]$$

$$= -2.17 \times 10^{-4} \text{ in}$$

The total radial interference is

$$1.20 \times 10^{-3} + 2.17 \times 10^{-4} = \boxed{1.42 \times 10^{-3}}$$

5. Proceed as in warmup 1. First, find the pitch diameters.

$$\frac{d_1}{2} + \frac{d_2}{2} = 15 \text{ in}$$

$$\frac{d_1}{d_2} = \frac{250}{83.33}$$

Solving these simultaneously,

$$d_1 = 22.5 \text{ in}$$

$$d_2 = 7.5 \text{ in}$$

The pitch line velocity is

$$v = \frac{\pi(\text{rpm})d}{12} = \frac{\pi(7.5)(250)}{12}$$

$$= 490.9 \text{ ft/min}$$

For the pinion, assume $Y = 0.3$, and $w = 1$ in. Use $\sigma_{max} = 30,000$ psi.

The Barth speed factor is

$$k_d = \frac{600}{600 + 490.9} = 0.55$$

$$P = \frac{k_d \sigma w Y v}{(\text{hp})(33,000)}$$

$$= \frac{(0.55)(30,000)(1)(0.3)(490.9)}{(40)(33,000)}$$

$$= 1.84$$

Choose $P = 2$. The numbers of teeth are

$$N_{\text{pinion}} = Pd = (2)(7.5) = 15$$
$$N_{\text{gear}} = (2)(22.5) = 45$$

The actual values of Y are

$$Y_{\text{pinion}} = 0.290$$
$$Y_{\text{gear}} \approx 0.401$$

The widths are

$$w = \frac{(33{,}000)P(\text{hp})}{v\sigma Y k_d}$$

$$w_{\text{pinion}} = \frac{(33{,}000)(2)(40)}{(490.9)(30{,}000)(0.290)(0.55)}$$
$$= 1.12 \text{ in}$$

Assume $\sigma_{\text{maximum}} = 50{,}000$ and a factor of safety of 3 for the cast steel gear.

$$w_{\text{gear}} = \frac{(33{,}000)(2)(40)}{(490.9)\left(\dfrac{50{,}000}{3}\right)(0.401)(0.55)}$$

$$= \boxed{1.46 \text{ in}}$$

Note: Practical gear design requires the face width to be 3–5 times π/P. Thus, these theoretical widths are too small.

6.

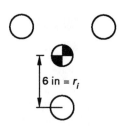

The centroid of the rivet group is

$$\left(\tfrac{2}{3}\right)(9) = 6 \text{ in} \quad \text{[from each rivet]}$$

The vertical shear load carried by each rivet is

$$F_{\text{v}} = \frac{P}{3}$$

For each rivet, the resisting moment of inertia is

$$J = A r_i^2 = \pi r^2 r_i^2$$
$$= \pi \left(\frac{0.75}{2}\right)^2 (6)^2$$
$$= 15.9 \text{ in}^4$$

The applied moment is $20P$.

The area of each rivet is

$$A = \left(\frac{\pi}{4}\right) d^2 = \left(\frac{\pi}{4}\right)(0.75)^2$$
$$= 0.442 \text{ in}^2$$

The vertical shear stress is

$$\tau_{\text{v}} = \frac{F_{\text{v}}}{A} = \frac{P}{(3)(0.442)}$$
$$= 0.754p$$

The twisting shear stress in the right-most rivet (which, by inspection, is the highest stressed) is

$$\tau = \frac{Mr}{J} = \frac{\left(\dfrac{20P}{3}\right)6}{15.9}$$
$$= 2.516p$$

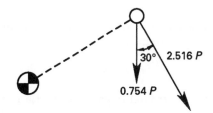

Combining the two stresses,

$$\tau_{\text{max}} = P\sqrt{\begin{array}{l}[(2.516)\sin 30°]^2 \\ + [0.754 + (2.516)\cos 30°]^2\end{array}}$$

$$= 3.191P$$

Keeping $\tau_{\text{max}} < 15{,}000$ psi,

$$P = \frac{15{,}000}{3.191} = \boxed{4700 \text{ lbf}}$$

7. Assume the weld throat thickness is t.

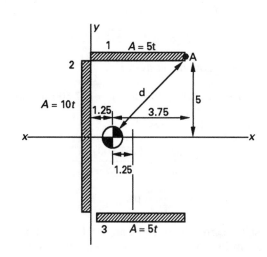

By inspection, $\bar{y} = 0$.

$$A_1 = 5t$$
$$\bar{x}_1 = 2.5$$
$$A_2 = 10t$$
$$\bar{x}_2 = 0$$
$$A_3 = 5t$$
$$\bar{x}_3 = 2.5$$
$$\bar{x}_c = \frac{5t(2.5) + 10t(0) + 5t(2.5)}{5t + 10t + 5t}$$
$$= 1.25 \text{ in}$$

The controidal moment of inertia in the x direction is

$$I_x = \frac{t(10)^3}{12} + (2)\left[\frac{5t^3}{12} + 5t(5)^2\right]$$
$$= 333.33t + 0.833t^3$$

Since t will be small (probably less than 0.5 in), the t^3 term may be omitted. So,

$$I_x = 333.33t$$

Similarly, the centroidal moment of inertia in the y direction is

$$I_y = \frac{10t^3}{12} + 10t(1.25)^2 + (2)\left[\frac{t(5)^3}{12} + (5t)(1.25)^2\right]$$
$$= 0.833t^3 + 15.625t + 20.833t + 15.625t$$
$$= 0.833t^3 + 52.08t$$
$$\approx 52.08t \text{ in}^4$$

The polar moment of inertia is

$$J = I_x + I_y = 385.4t \text{ in}^4$$

The maximum shear will occur at point A since it is farthest away.

$$d = \sqrt{(3.75)^2 + (5)^2} = 6.25 \text{ in}$$

The applied moment is $m = Fx$.

$$M = (10,000)(12 + 3.75) = 157,500 \text{ in-lbf}$$

The torsional shear stress is

$$\tau = \frac{Mc}{J} = \frac{(157,500)(6.25)}{385.4t}$$
$$= \frac{2554.2}{t} \text{ psi}$$

The shear stress is at right angles to the line d. The stress can be divided into x and y components.

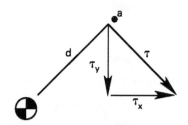

$$\tau_y = \left(\frac{3.75}{6.25}\right)\left(\frac{2554.2}{t}\right) = \frac{1532.5}{t} \text{ psi}$$
$$\tau_x = \left(\frac{5}{6.25}\right)\left(\frac{2554.2}{t}\right) = \frac{2043.4}{t} \text{ psi}$$

In addition, the vertical load of 10,000 lbf must be supported. The vertical shear stress is

$$A = 10t + 5t + 5t = 20t$$
$$\tau_y = \frac{10,000}{20t} = \frac{500}{t} \text{ psi}$$

The resultant shear stress is

$$\tau = \sqrt{\left(\frac{2043.4}{t}\right)^2 + \left(\frac{1532.5 + 500}{t}\right)^2}$$
$$= \frac{2882.1}{t} \text{ psi}$$

Assuming an allowable load of 8000 psi for an unshielded weld in shear, the required throat thickness is

$$\frac{2882.1}{t} = 8000$$
$$t = 0.360 \text{ in}$$
$$w = \frac{0.360}{0.707}$$
$$= \boxed{0.51 \text{ in} \quad [\text{say } \tfrac{1}{2} \text{ in}]}$$

8.

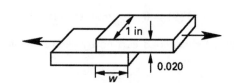

The aluminum strength per inch of bond is

$$S = (15,000)(1)(0.020) = 300 \text{ lbf}$$

The width of bond required is

$$300 = \left(\frac{1500}{2}\right)(1)(w)$$
$$w = \boxed{0.4 \text{ in}}$$

9. The centroidal moment of inertia is

$$\left(\frac{\pi}{4}\right) r^4 = \left(\frac{\pi}{4}\right)(1)^4 = 0.7854 \text{ in}^4$$

Although the stress concentration is probably the greatest at the fillet, it is difficult to predict where the greatest stress will occur. Therefore, by common convention, the stress is assumed to occur at the shoulder.

Assuming direct shear is low by comparison, the moment is

$$M = (2500)(4 \text{ in}) = 10,000 \text{ in-lbf}$$

$$\frac{D}{d} = \frac{3}{2} = 1.5$$

$$\frac{r}{d} = \frac{\frac{5}{16}}{2} = 0.156$$

The stress concentration factor is $k \approx 1.5$. The bending stress is

$$\sigma = k\left(\frac{Mc}{T}\right) = (1.5)\left[\frac{(10,000)(1)}{0.7854}\right]$$

$$= \boxed{19,099 \text{ psi}}$$

10. Neglect the expansion of the shaft diameter, which is small anyway. (This is a conservative assumption.)

Refer to *Roark's Formulas for Stress and Strain* (Young, 1989). The change in inner radius is

$$\Delta R_i = \left(\tfrac{1}{4}\right)\left(\frac{\rho\omega^2}{386.4}\right)\left(\frac{R_i}{E}\right)$$

$$\times \left[(3+\mu)R^2 + (1-\mu)R_i^2\right] \quad \text{[in inches]}$$

Poisson's ratio is

$$\mu = 0.3$$

The steel density is

$$\rho = 0.283 \text{ lbm/in}^3$$

$$\omega = \frac{2\pi(3500)}{60} = 366.5 \text{ rad/sec}$$

$$E = 2.9 \times 10^7 \text{ psi}$$

The outer radius is

$$R = 8 \text{ in}$$

The inner radius is

$$R_i = 1 \text{ in}$$

$$\Delta R_i = \left(\tfrac{1}{4}\right)\left[\frac{(0.283)(366.5)^2}{386.4}\right]\left(\frac{1}{2.9 \times 10^7}\right)$$

$$\times \left[(3+0.3)(8)^2 + (1-0.3)(1)^2\right]$$

$$= 1.80 \times 10^{-4} \text{ in}$$

Since both the shaft and the disk are steel, the interference required to obtain a residual pressure of 1250 psi is

$$I_{\text{diametral}} = \Delta D_i = \left[\frac{(4)(1)(1250)}{2.9 \times 10^7}\right]\left[\frac{1}{1-\left(\frac{1}{8}\right)^2}\right]$$

$$= 1.75 \times 10^{-4} \text{ in}$$

$$I_{\text{radial}} = \left(\tfrac{1}{2}\right) I_{\text{diametral}}$$

$$= 8.76 \times 10^{-5} \text{ in}$$

The required initial interference is

$$I_{\text{radial}} = 1.80 \times 10^{-4} + 8.76 \times 10^{-5}$$

$$= \boxed{2.68 \times 10^{-4} \text{ in}}$$

11.

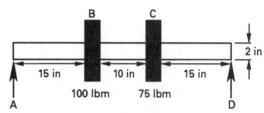

$$I_{\text{shaft}} = \tfrac{1}{4}\pi r^4 = (0.25)\pi(1)^4$$

$$= 0.7854 \text{ in}^4$$

Calculate the deflections due to the 100-lbm load.

$$\delta_{\text{B},100} = \frac{Fa^2b^2}{3EIL}$$

$$= \frac{(100)(15)^2(25)^2}{(3)(3 \times 10^7)(0.7854)(40)}$$

$$= 4.97 \times 10^{-3} \text{ in}$$

$$\delta_{\text{C},100} = \left(\frac{Fbx}{6EIL}\right)(L^2 - b^2 - x^2)$$

$$= \left[\frac{(100)(15)(15)}{(6)(3 \times 10^7)(0.7854)(40)}\right]$$

$$\times \left[(40)^2 - (15)^2 - (15)^2\right]$$

$$= 4.58 \times 10^{-3} \text{ in}$$

Alternate method:

Alternatively, the values of a and b can be reversed to calculate the deflection.

$$\delta_{C,100} = \frac{Fb}{6EIL} = \left[\left(\frac{L}{b}\right)(x-a)^3 + (L^2-b^2)x - x^3\right]$$

$$= \left[\frac{(100)(25)}{(6)(3\times10^7)(0.7854)(40)}\right]$$

$$\times \left[\left(\frac{40}{25}\right)(25-15)^3 + \left[(40)^2-(25)^2\right](25) - (25)^3\right]$$

$$= 4.58\times10^{-3} \text{ in}$$

Calculate the deflections due to the 75-lbm load.

$$\delta_{C,75} = \frac{(75)(25)^2(15)^2}{(3)(3\times10^7)(0.7854)(40)}$$

$$= 3.73\times10^{-3} \text{ in}$$

$$\delta_{B,75} = \frac{(75)(15)(15)}{(6)(3\times10^7)(0.7854)(40)}$$

$$\times \left[(40)^2-(15)^2-(15)^2\right]$$

$$= 3.43\times10^{-3} \text{ in}$$

The total deflections are

$$\delta_B = 4.97\times10^{-3} + 3.43\times10^{-3} \text{ in}$$
$$= 8.4\times10^{-3} \text{ in}$$
$$\delta_C = 4.58\times10^{-3} + 3.73\times10^{-3}$$
$$= 8.31\times10^{-3} \text{ in}$$

The critical shaft speed is

$$f = \frac{1}{2\pi}\sqrt{\frac{g\sum w_i\delta_i}{\sum w_i\delta_i^2}}$$

$$= \frac{1}{2\pi}\sqrt{\frac{(386)\left[(100)(8.4\times10^{-3})+(75)(8.31\times10^{-3})\right]}{(100)(8.4\times10^{-3})^2+(75)(8.31\times10^{-3})^2}}$$

$$= \boxed{34.2 \text{ Hz} = 2052 \text{ rpm}}$$

12.
$$N_S = 24$$
$$N_P = 40$$
$$N_R = 104$$
$$\omega_S = +50 \text{ rpm}$$
$$\omega_R = 0$$

$$TV = -\frac{N_R}{N_S} = -\frac{104}{24} = -4.333$$
$$\omega_S = (TV)(\omega_r) + \omega_C(1-TV)$$
$$\omega_C = \frac{+50}{1-(-4.333)}$$
$$= \boxed{+9.376 \text{ rpm} \quad [\text{clockwise}]}$$

13.
$$\omega_R = \omega_A\left(\frac{-N_A}{N_R}\right) = (-100)\left(\frac{-30}{100}\right)$$
$$= 30 \text{ rpm} \quad [\text{clockwise}]$$
$$TV = \frac{-N_R}{N_S} = -\frac{80}{40}$$
$$= -2$$
$$\omega_S = (TV)(\omega_R) + \omega_C(1-TV)$$
$$= (-2)(+30) + (60)\left[1-(-2)\right]$$
$$= \boxed{+120 \text{ rpm} [\text{clockwise}]}$$

14.
$$N_S = (10)(5) = 50$$
$$N_P = (10)(2.5) = 25$$
$$N_R = (10)(10) = 100$$
$$\omega_R = +1500 \text{ rpm}$$
$$\omega_s = 0$$
$$TV = -\frac{N_R}{N_S} = \frac{-100}{50}$$
$$= -2$$
$$\omega_S = (TV)(\omega_R) + \omega_C(1-TV)$$
$$0 = (-2)(1500) + \omega_C\left[1-(-2)\right]$$
$$\omega_C = +1000 \text{ rpm}$$
$$T_{input} = \frac{(33,000)(\text{hp})}{(\text{rpm})(2\pi)} = \frac{(33,000)(15)}{(1500)(2\pi)}$$
$$= \boxed{52.52 \text{ ft-lbf}}$$
$$T_{output} = \frac{(33,000)(15)}{(1000)(2\pi)}$$
$$= \boxed{78.78 \text{ ft-lbf}}$$

Timed

1.
$$\mu = 1.184 \times 10^{-6} \text{ reyns}$$

$$n = \frac{1200 \text{ rpm}}{60} = 20 \text{ rps}$$

$$w = 880 \text{ lbf}$$

$$\frac{c_d}{d} = \frac{2c_r}{2r} = \frac{c_r}{r}$$

$$\frac{c_r}{r} = \frac{\text{radial clearance}}{\text{journal radius}} = \frac{1}{1000}$$

$$r = 1.5 \text{ in}$$

$$\ell = 3.5 \text{ in}$$

The unit load is

$$p = \frac{w}{2r\ell} = \frac{880}{(2)(1.5)(3.5)}$$

$$= 83.81 \text{ psi}$$

The bearing characteristic number is

$$S = \left(\frac{r}{c}\right)^2 \left[\frac{\mu N}{p}\right]$$

$$= (1000)^2 \left[\frac{(1.184 \times 10^{-6})(20)}{83.81}\right]$$

$$= 0.283$$

$$\frac{\ell}{d} = \frac{\ell}{2r} = \frac{3.5}{3}$$

$$= 1.17$$

The friction variable is 6, so the coefficient of friction is

$$f = (6)\left(\frac{c}{r}\right) = \frac{6}{1000}$$

$$= 0.006$$

The friction torque is

$$T = fwr = (0.006)(880)(1.5)$$

$$= 7.92 \text{ in-lbf}$$

The friction horsepower is

$$\text{hp} = \frac{(7.92 \text{ in-lbf})\left(20 \frac{1}{\text{sec}}\right)(2\pi)}{\left(12 \frac{\text{in}}{\text{ft}}\right)\left(550 \frac{\text{ft-lbf}}{\text{hp-sec}}\right)}$$

$$= 0.151 \text{ hp}$$

The film thickness variable is 0.64. The film thickness is

$$h_o = (1.5)\left(\frac{1}{1000}\right)(0.64) = \boxed{0.00096 \text{ in}}$$

Because this is out of the recommended region, the load is probably too much for the bearing.

2. Because of their hardness, assume

$$E = 3.0 \times 10^7 \text{ for the bolts}$$

$$= 2.9 \times 10^7 \text{ for the vessel}$$

The bolt stiffness is

$$k_b = \frac{AE}{L} = \frac{(6)\left(\frac{\pi}{4}\right)\left(\frac{3}{8}\right)^2 (3 \times 10^7)}{\frac{3}{8} + \frac{3}{4}}$$

$$= 1.767 \times 10^7 \text{ lbf/in}$$

The plate/vessel contact area is an annulus with inside and outside diameters of 8.5 in and 11.5 in, respectively.

$$\text{area} = \left(\frac{\pi}{4}\right)\left[(11.5)^2 - (8.5)^2\right] = 47.12 \text{ in}^2$$

The vessel/plate stiffness is

$$k_m = \frac{AE}{L} = \frac{(47.12)(2.9 \times 10^7)}{\frac{3}{8} + \frac{3}{4}}$$

$$= 1.215 \times 10^9$$

Let x be the decimal portion of the pressure load, p, taken by the bolts. The increase in the bolt length is

$$\delta_b = \frac{xp}{k_b}$$

The plate/vessel deformation is

$$\delta_m = \frac{(1-x)p}{k_m}$$

$$\delta_b = \delta_m$$

$$\frac{x}{1.767 \times 10^7} = \frac{1-x}{1.215 \times 10^9}$$

$$x = 0.0143 \ (1.43\%)$$

The end plate area exposed to pressure is

$$\left(\frac{\pi}{4}\right)(8.5)^2 = 56.75 \text{ in}^2$$

The forces on the end plate are

$$F_{\max} = pA = (350)(56.75)$$

$$= 19{,}860 \text{ lbf}$$

$$F_{\min} = (50)(56.75) = 2838 \text{ lbf}$$

The stress area in the bolt is 0.0876 in^2.

The stresses in each bolt are

$$\sigma_{max} = \frac{3700 + \frac{(0.0143)(19,860)}{6}}{0.0876}$$

$$= 42,778 \text{ psi}$$

$$\sigma_{min} = \frac{3700 + \frac{(0.0143)(2838)}{6}}{0.0876}$$

$$= 42,315 \text{ psi}$$

The mean stress is

$$\sigma_{mean} = \left(\tfrac{1}{2}\right)(42,778 + 42,315) = 42,547 \text{ psi}$$

Using the thread concentration factor of 2, the alternating stress is

$$\sigma_{alt} = \left(\tfrac{1}{2}\right)(2)(42,778 - 42,315)$$

$$= 463 \text{ psi}$$

For the bolt material,

$$S'_e = 0.5 S_{ut} = (0.5)(110,000)$$

$$= 55,000 \text{ psi}$$

Derating factors from other sources are

$$k_a = 0.72 \quad \text{[surface finish]}$$
$$k_b = 0.85 \quad \text{[size]}$$
$$k_c = 0.90 \quad \text{[reliability]}$$

$$S_e = (0.72)(0.85)(0.90)(55,000)$$

$$= 30,294 \text{ psi}$$

Plotting the point on the modified Goodman diagram,

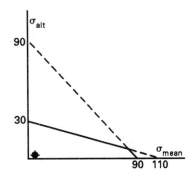

It is apparent that the factor of safety is quite high, and the bolts are in no danger of failing.

The stress in the flange/vessel is

$$\sigma_{min} = \frac{(6)(3700) - (1 - 0.0143)(19,860)}{47.12}$$

$$= 55.69 \text{ psi}$$

$$\sigma_{max} = \frac{(6)(3700) - (1 - 0.0143)(2838)}{47.12}$$

$$= \boxed{411.8 \text{ psi}}$$

These stresses are very low. The plate stresses should also be checked.

3. Assume the diameters are mean diameters. The spring indexes are

$$C_{inner} = \frac{D}{d} = \frac{1.5}{0.177}$$

$$= 8.475$$

$$C_{outer} = \frac{2.0}{0.2253} = 8.877$$

For oil-hardened steel, $G = 11.5 \times 10^6$ psi.

Assume the number of coils is the number of total coils. The number of active coils is 2 less than the total since the springs have squared and ground ends.

$$N_{a,inner} = 12.75 - 2 = 10.75$$

$$N_{a,outer} = 10.25 - 2 = 8.25$$

The spring rates (stiffness) are

$$k_{inner} = \frac{(11.5 \times 10^6)(0.177)}{(8)(8.475)^3(10.75)}$$

$$= 38.88 \text{ lbf/in}$$

$$k_{outer} = \frac{(11.5 \times 10^6)(0.2253)}{(8)(8.877)^3(8.25)}$$

$$= 56.12 \text{ lbf/in}$$

Since the inner spring is longer by $4.5 - 3.75 = 0.75$ in, the inner spring will absorb an initial force of

$$(38.88)(0.75) = 29.16 \text{ lbf}$$

The remaining $150 - 29.16 = 120.84$ lbf will be shared by both springs.

The composite spring constant is

$$38.88 + 56.12 = 95.0 \text{ lbf/in}$$

The deflection of the outer spring is

$$\delta_{outer} = \frac{120.84}{95} = 1.272 \text{ in}$$

The total deflection of the inner spring is

$$\delta_{\text{inner}} = 0.75 + 1.272 = \boxed{2.022 \text{ in}}$$

The total force exerted by the inner spring is

$$F = k\delta = (38.88)(2.022) = 78.62 \text{ lbf}$$

The Wahl factor is

$$W = \frac{(4)(8.475) - 1}{(4)(8.475) - 4} + \frac{0.615}{8.475}$$

$$= 1.173$$

The shear stress is

$$\tau_{\text{inner}} = \frac{(8)(8.475)(78.62)(1.173)}{\pi(0.177)^2}$$

$$= 63{,}528 \text{ psi}$$

According to the maximum shear stress theory,

$$\tau_{\text{max}} = 0.5 S_{yt}$$

Unfortunately, the yield strength is not given. For oil-tempered steel,

$$S_{ut} \approx 215 - \left(\frac{0.177 - 0.15}{0.20 - 0.15} \right)(215 - 195)$$

$$= 204 \text{ ksi}$$

$$S_{yt} \approx 0.75 S_{ut} = (0.75)(204)$$

$$= 153 \text{ ksi}$$

Then,

$$\tau_{\text{max}} \approx (0.5)(153) = 76.5 \text{ ksi}$$

The 0.577 multiplier factor is not used because it was derived from the distortion energy theory, not from the maximum shear stress theory.

The factor of safety is

$$FS = \frac{76.5}{63.5} = \boxed{1.2}$$

The answer will vary depending on the method of calculating S_{yt}.

The inner and outer springs should be wound with opposite direction helixes. This will minimize resonance and prevent coils from one spring entering the other spring's gaps.

4. This is an automobile differential, so both output shafts must turn in the same direction.

This is identical to the following gear set.

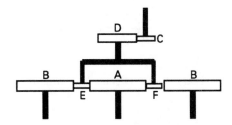

The difference between this and a conventional epicyclic gear train is that the ring gear is replaced with two external gears, B.

$$\omega_C = +600 \text{ rpm}$$

$$\omega_D = \left(\frac{-18}{54} \right)(600) = -200 \text{ rpm}$$

$$\omega_A = -50 \text{ rpm}$$

$$TV = \frac{-N_B}{N_A} = \frac{-30}{30} = -1$$

TV is clearly negative. Look at the actual gear set (not the equivalent gear set above). Hold gear D stationary. Turn gear A. Then gear B moves in the opposite direction of gear A.

$$\omega_A = (TV)\omega_B + \omega_D(1 - TV)$$

$$-50 = (-1)\omega_B - (200)[1 - (-1)]$$

$$\omega_B = \boxed{-350 \text{ rpm}}$$

5. $$\omega_R = 0$$

$$\omega_S = 1000 \text{ rpm}$$

$$\omega_C = \frac{\omega_s}{3} = 333.3 \text{ rpm}$$

$$1000 = (TV)0 = (333.3)(1 - TV)$$

$$TV = -2$$

$$2 = \frac{N_R}{N_S}$$

Assume $N_S = 40$ and $N_R = 80$. Since $P = 10$,

$$D_S = 4$$

$$D_R = 8$$

$$D_P = 2$$

Check: $8 = 4 + (2)(2)$ [ok]

$$\frac{20}{40} = -\frac{1000 - 333.3}{\omega_P - 333.3}$$

$$\omega_P = \boxed{-1000 \text{ rpm}}$$

6. First, simplify the problem. Input 2 causes gear D-E to turn.

$$\omega_D = (-75)\left(\frac{-28}{32}\right) = 65.625 \text{ rpm} \quad [\text{CW}]$$

$$= 65.625 \text{ rpm}$$

step 1: Gears A and D have the same center.

A	D

step 2:

	A	D
1	ω_{arm}	ω_{arm}
2		
3		

step 3: Since all speeds are known, choose gear D arbitrarily as the "unknown."

	A	D
1	ω_{arm}	ω_{arm}
2		
3		ω_D

step 4:

	A	D
1	ω_{arm}	ω_{arm}
2		$\omega_D - \omega_{\text{arm}}$
3		ω_D

step 5: We want the ratio ω_A/ω_D. The transmission path D→A is a compound mesh.

$$\omega_A = \omega_D\left(\frac{N_B N_D}{N_A N_C}\right) = \omega_D\left[\frac{(63)(56)}{(68)N_C}\right]$$

$$= \frac{51.8823\,\omega_D}{N_C}$$

Put $\left(\frac{51.8823}{N_C}\right)(\omega_D - \omega_{\text{arm}})$ into column A.

	A	D
1	ω_{arm}	ω_{arm}
2	$\left(\frac{51.8823}{N_C}\right)(\omega_D - \omega_{\text{arm}})$	$\omega_D - \omega_{\text{arm}}$
3		ω_D

step 6: Insert the known values into ω.

	A	D
1	150	150
2	$\left(\frac{51.8823}{N_C}\right)(65.625 - 150)$	$65.625 - 150$
3	-250	65.625

step 7:

$$\text{row } 1 + \text{row } 2 = \text{row } 3$$

$$150 + \left(\frac{51.8823}{N_C}\right)(65.625 - 150) = -250$$

Solving for N_C gives

$$\boxed{N_C = 10.94 \approx \boxed{11}}$$

7. (a) The output speed is

$$(1800)\left(\frac{50}{25}\right)\left(\frac{60}{20}\right) = \boxed{10{,}800 \text{ rpm}}$$

(b)
$$T_4 = \frac{(63{,}025)(50)(0.98)^2}{10{,}800}$$

$$= \boxed{280 \text{ in-lbf}}$$

(c) The shaft torque is

$$T_{2\text{-}3} = \frac{(63{,}025)(50)(0.98)}{3600}$$

$$= 857.84 \text{ in-lbf}$$

For cold-drawn 1045 steel,

$$S_{ys} = 69{,}000 \text{ psi}$$
$$\tau_{\text{max}} = \left(\tfrac{1}{2}\right)(69{,}000)$$
$$= 34{,}500 \text{ psi}$$
$$P = 5 = \text{no. teeth}/d_p$$

For gear 3, $d_p = 60/5 = 12$-in-diameter. The pitch circle velocity is

$$v_r = \frac{(3600)\pi(12)}{12} = 11{,}310 \text{ ft/min}$$

$$F_t = \frac{(33{,}000)(0.98)(50)}{11{,}310} = 143 \text{ lbf}$$

$$\phi_t = \arctan\left(\frac{\tan 20°}{\cos 25°}\right) = 21.88°$$

$$F_r = (143)\tan 21.88° = 57.4 \text{ lbf}$$

$$F_a = (143)\tan 25° = 66.7 \text{ lbf}$$

The resulting bending moment on the shaft is

$$M = (57.4 \text{ lbf})(4 \text{ in}) = 229.6 \text{ in-lbf}$$

For a steady load, $k_m = 1.5$, and $k_t = 1.0$. With a factor of safety of 3,

$$d^3 = \left[\frac{(16)(3)}{\pi \left(\frac{1}{2}\right)(69,000)} \right]$$

$$\times \sqrt{[(1.5)(229.6)]^2 + [(1)(857.84)]^2}$$

$$= 0.40938 \text{ in}^3$$

$$d = 0.74 \text{ in} \quad \begin{bmatrix} \text{value will depend} \\ \text{on choice of } \tau_{max} \end{bmatrix}$$

Check the axial stresses.

$$\sigma_{axial} = \frac{F_a}{A} = \frac{66.7 \text{ lbf}}{\left(\frac{\pi}{4}\right)(0.74 \text{ in})^2}$$

$$= 155 \text{ psi}$$

This stress is much less than the yield strength (69,000 psi), thus it is insignificant in this problem. The minimum diameter, then, should be 0.74 in. For transmission shafting,

> the standard diameter of 15/16 in = 0.9375 in would be adequate

8. $S_{ut} = 30,000$ psi. Choose a safety factor of 10. $\mu = 0.27$. For tangential stress,

$$\omega = \sqrt{\frac{(4)(386)\left(\frac{30,000}{10}\right)}{(0.26)[(3.27)(10)^2 + (1-0.27)(2)^2]}}$$

$$= 232.4 \text{ rad/sec}$$

$$n = \frac{(60)(232.4)}{2\pi} = 2219 \text{ rpm}$$

For radial stress,

$$n = \left(\frac{30}{\pi}\right) \sqrt{\left(\frac{8}{3.27}\right)(3000)\left[\frac{386}{(0.26)(10-2)^2}\right]}$$

$$= 3940 \text{ rpm}$$

The maximum safe speed, then, is $\boxed{2219 \text{ rpm}}$

9. (b) $\omega = \left(\frac{120 \text{ rpm}}{60}\right)(2\pi) = 12.57$ rad/sec

At $\theta = 60°$,

$$t = \frac{\left(\frac{60}{180}\right)(\pi)}{12.57} = 0.08331 \text{ sec}$$

Since the acceleration is constant,

$$\frac{d^2 x}{dt^2} = a$$

$$x = \tfrac{1}{2}at^2$$

$$a = \frac{(2)(0.5)}{(0.08331)^2}$$

$$= \boxed{144.08 \text{ in/sec}^2}$$

(c) At $\theta = 60°$, since $a = dv/dt$,

$$v = at = \left(144.08 \frac{\text{in}}{\text{sec}^2}\right)(0.08331)$$

$$= 12 \text{ in/sec}$$

The cam follower returns to rest at $\theta = 150°$, so with constant acceleration, starting at $v_i = 12$ in/sec at $t = 0$, $v_f = 0$.

$$t_f = \frac{\left(\frac{90}{180}\right)\pi}{12.57}$$

$$= 0.1250 \text{ sec} \quad \begin{bmatrix} \text{time between} \\ \theta = 60° \text{ and} \\ \theta = 150° \end{bmatrix}$$

$$v_f - v_i = a(t_f - t_i)$$

$$12 - 0 = a(0.1250 - 0)$$

$$a = \boxed{-96 \text{ in/sec}^2}$$

(a) The distance moved by the cam follower during the $60° \le \theta \le 150°$ is

$$x = \tfrac{1}{2}at^2 = \left(\tfrac{1}{2}\right)(96)(0.1250)^2$$

$$= 0.75 \text{ in}$$

The total follower distance is

$$0.5 \text{ in} + 0.75 \text{ in} = \boxed{1.25 \text{ in}}$$

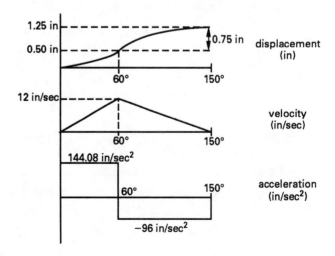

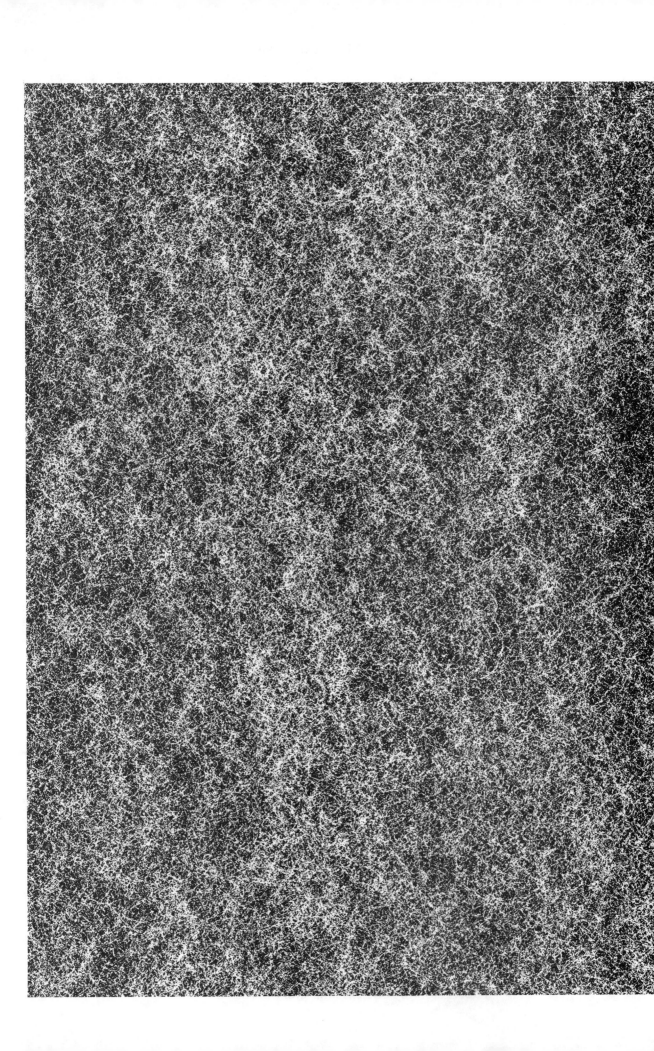

DYNAMICS

Warmups

1.

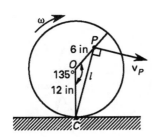

$$\omega = \left(\frac{\text{rev}}{\text{sec}}\right)(2\pi)$$

$$= \frac{\left(28\ \frac{\text{mi}}{\text{hr}}\right)\left(5280\ \frac{\text{ft}}{\text{mi}}\right)\left(12\ \frac{\text{in}}{\text{ft}}\right)(2\pi)}{\pi(24\ \text{in})\left(3600\ \frac{\text{sec}}{\text{hr}}\right)}$$

$$= 41.07\ \text{rad/sec}$$

Find distance l.

$$l^2 = (12)^2 + (6)^2 - (2)(12)(6)\cos 135°$$

$$l = 16.79\ \text{in}$$

$$v_P = \omega l = \left(41.07\ \frac{\text{rad}}{\text{sec}}\right)(16.79\ \text{in})$$

$$= 689.6\ \text{in/sec}$$

$$= \boxed{57.46\ \text{ft/sec}}$$

2.

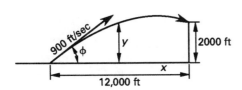

$$x = (v_o \cos\phi)t$$

$$= (900\cos\phi)t$$

$$t = \frac{12,000}{900\cos\phi}$$

$$= \frac{13.33}{\cos\phi}$$

$$y = (v_o \sin\phi)t - \tfrac{1}{2}gt^2$$

$$= (900\sin\phi)t - \left(\tfrac{1}{2}\right)(32.2)t^2$$

Substituting $y = 2000$ and $t = 13.33/\cos\phi$,

$$2000 = \frac{900\sin\phi(13.33)}{\cos\phi}$$

$$- (16.1)\left(\frac{13.33}{\cos\phi}\right)^2$$

$$1 = 6\tan\phi - \frac{1.43}{\cos^2\phi}$$

$$\frac{1}{\cos^2\phi} = 1 + \tan^2\phi$$

$$\tan^2\phi - 4.2\tan\phi + 1.7 = 0$$

$$\tan\phi = \left\{\begin{matrix} 3.747 \\ 0.453 \end{matrix}\right\}$$

$$\phi = \boxed{\left\{\begin{matrix} 75.06° \\ 24.37° \end{matrix}\right\}}$$

Concentrates

1. The static deflection is caused by the 300 lbm magnet.

$$\delta_{\text{st}} = \left(\frac{300\ \text{lbm}}{1000\ \frac{\text{lbf}}{\text{in}}}\right)\left(\frac{g}{g_c}\right) = 0.3\ \text{in}$$

$$f_{\text{nat}} = \frac{1}{2\pi}\sqrt{\frac{2}{\delta_{\text{st}}}} = \frac{1}{2\pi}\sqrt{\frac{386}{0.3}}$$

$$= \boxed{5.71\ \text{Hz}}$$

Minimum tension occurs at the upper limit of travel. The decrease in tension at that point is the same as the increase at the lower limit caused by the 200 lbm of scrap.

$$F_{\text{min}} = 300 - 200 = \boxed{100\ \text{lbf}}$$

2. If the springs reduce the force from 25 lbf to 3 lbf, the transmissivity is

$$\text{TR} = \frac{-3}{25} = -0.12$$

(TR and β are negative whenever they are < 1.)

The angular forcing frequency is

$$\omega_f = 2\pi f = \left(2\pi\ \frac{\text{rad}}{\text{sec}}\right)\left(\frac{1200\ \frac{\text{rev}}{\text{min}}}{60\ \frac{\text{sec}}{\text{min}}}\right)$$

$$= 125.66\ \text{rad/sec}$$

$$\text{TR} = \frac{1}{1 - \left(\frac{\omega_f}{\omega}\right)^2}$$

$$-0.12 = \frac{1}{1 - \left(\frac{125.66}{\omega}\right)^2}$$

$$\omega = 41.13 \text{ rad/sec}$$

For the spring-mounted mass,

$$\omega = \sqrt{\frac{k}{m}}$$

$$41.13 = \sqrt{\frac{386k}{800}}$$

$$k = 3506.1 \text{ lbf/in}$$

$$k_{\text{spring}} = \frac{k}{4} = 876.5 \text{ lbf/in}$$

The static deflection is

$$\delta_{\text{st}} = \frac{\text{weight}}{k} = \frac{800 \text{ lbf}}{3506.1 \; \frac{\text{lbf}}{\text{in}}}$$

$$= 0.228 \text{ in}$$

The added deflection due to the oscillatory force is

$$\frac{3 \text{ lbf}}{3506.1 \; \frac{\text{lbf}}{\text{in}}} = \boxed{0.000856 \text{ in}}$$

3.

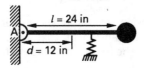

The moment of inertia of the arm about point A is

$$J_{\text{A,arm}} = \left(\frac{1}{12}\right) ml^2 + md^2$$

$$= \left(\frac{1}{12}\right)\left(\frac{5}{386}\right)(24)^2 + \left(\frac{5}{386}\right)(12)^2$$

$$= 2.487 \text{ lbf-in-sec}^2$$

The moment of inertia of the concentrated load is

$$J_{\text{A,load}} = ml^2 = \left(\frac{3}{386}\right)(24)^2$$

$$= 4.477 \text{ lbf-in-sec}^2$$

$$J_{\text{total}} = 2.487 + 4.477 = 6.964 \text{ lbf-in-sec}^2$$

The moment causing deflection (which is resisted in total by the spring moment) is found by summing moments about point A. Taking clockwise moments as positive,

$$\sum M_A: \; (3)(24) + (5)(12) - M_{\text{spring}} = 0$$

$$M_{\text{spring}} = 132 \text{ in-lbf}$$

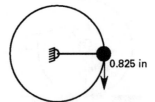

The deflection at the tip is

$$(0.55)\left(\frac{24}{16}\right) = 0.825 \text{ in}$$

This deflection in radians is

$$\frac{(0.825 \text{ in})\left(2\pi \; \frac{\text{rad}}{\text{rev}}\right)}{(2\pi)\left(24 \; \frac{\text{in}}{\text{rev}}\right)} = 0.0344 \text{ rad}$$

The angular stiffness of the system is the moment that would cause a 1-rad deflection.

$$k_R = \frac{132 \text{ in-lbf}}{0.0344 \text{ rad}} = 3837.2 \text{ in-lbf/rad}$$

The natural frequency is

$$f = \frac{1}{2\pi}\sqrt{\frac{k_R}{J}} = \frac{1}{2\pi}\sqrt{\frac{3837.2}{6.964}}$$

$$= \boxed{3.74 \text{ Hz}}$$

4.

$$M = \frac{1 \text{ oz}}{\left(16 \; \frac{\text{oz}}{\text{lbm}}\right)\left(386 \; \frac{\text{in}}{\text{sec}^2}\right)}$$

$$= 0.000162 \text{ lbm-sec}^2/\text{in}$$

$$\omega_f = \frac{(2\pi \text{ rad/rev})\left(800 \; \frac{\text{rev}}{\text{min}}\right)}{60 \; \frac{\text{sec}}{\text{min}}}$$

$$= 83.78 \text{ rad/sec}$$

$$r = 5 \text{ in}$$

$$v_t = \omega r = \left(83.78 \; \frac{\text{rad}}{\text{sec}}\right)(5 \text{ in})$$

$$= 418.9 \text{ in/sec}$$

$$F_c = \text{centrifugal force} = \frac{m v_t^2}{r}$$

$$= \frac{(0.000162)(418.9)^2}{5}$$

$$= 5.69 \text{ lbf}$$

The natural frequency is

$$\omega = \sqrt{\frac{kg}{\text{weight}}} = \sqrt{\frac{(4)(1000)(386)}{50}}$$

$$= 175.7 \text{ rad/sec}$$

$$\frac{C}{C_{\text{crit}}} = \frac{1}{8}$$

$$\beta = \frac{1}{\sqrt{\left[1 - \left(\frac{\omega_f}{\omega}\right)^2\right]^2 + \left[(2)\left(\frac{C}{C_{\text{crit}}}\right)\left(\frac{\omega_f}{\omega}\right)\right]^2}}$$

$$= 1.28$$

The magnified excursion is

$$\delta = \frac{\beta F_c}{k} = \frac{(1.28)(5.69 \text{ lbf})}{(4)\left(1000 \dfrac{\text{lbf}}{\text{in}}\right)}$$

$$= \boxed{0.00182 \text{ in}}$$

Timed

1. The angular forcing frequency is

$$\omega_f = \frac{(1200)2\pi}{60} = 125.7 \text{ rad/sec}$$

The centrifugal force due to rotating this imbalance is

$$F_c = ma_N = mr\omega^2$$

$$= \left(\frac{3.6}{32.2}\right)\left(\frac{3}{12}\right)(125.7)^2$$

$$= 441.6 \text{ lbf}$$

The transmissivity is given as 0.05.

The natural frequency can be found from

$$\text{TR} = \frac{1}{\sqrt{\left[1 - \left(\frac{\omega_f}{\omega}\right)^2\right]^2}}$$

$$0.05 = \frac{1}{\left|1 - \left(\dfrac{125.7}{\omega}\right)^2\right|}$$

$$\omega = 27.43 \text{ rad/sec}$$

For a spring-mounted mass, the spring constant can be found from

$$\omega = \sqrt{\frac{k}{m}}$$

$$27.43 = \sqrt{\frac{k}{\dfrac{175}{386}}}$$

(using $g = 386$ in/sec^2)

$$k_{\text{total}} = 341.1 \text{ lbf/in}$$

Since the 4 corner springs are in parallel,

$$k_{\text{total}} = k_1 + k_2 + k_3 + k_4$$

$$k_{\text{each}} = \frac{341.1}{4}$$

$$= \boxed{85.28 \text{ lb/in}}$$

The amplitude of vibration (zero to peak) is

$$x = (\text{TR})\left(\frac{F}{k}\right) = (0.05)\left(\frac{441.6}{341.1}\right)$$

$$= \boxed{0.0647 \text{ in}}$$

2. The first plate under the load is a simple beam.

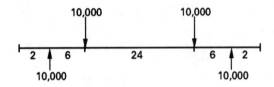

Assume $E = 2.9 \times 10^7$ psi.

$$I = \frac{bh^3}{12} = \frac{(30)\left(\frac{1}{2}\right)^3}{12}$$

$$= 0.3125 \text{ in}^4$$

The overhang contributes nothing to the rigidity. The deflection at the point of loading is

$$y = \left(\frac{Fx}{6EI}\right)\left[(3a)(L - a) - x^2\right]$$

$$= \left[\frac{(10,000)(6)}{(6)(2.9 \times 10^7)(0.3125)}\right]$$

$$\times \left[(3)(6)(36 - 6) - (6)^2\right]$$

$$y = 0.556 \text{ in}$$

The second plate is loaded exactly the same as the top plate, only upside down. Therefore, its deflection is also 0.556 in.

The total deflection of 8 plates is

$$y_{\text{total}} = (8)(0.556) = \boxed{4.448 \text{ in}}$$

This result assumes the yield point is not exceeded.

$$M_{\max} = Fa = (10{,}000)(6)$$
$$= 60{,}000 \text{ in-lbf}$$

$$\sigma_{\max} = \frac{M_{\max}c}{I}$$
$$= \frac{(60{,}000)(0.25)}{0.3125}$$
$$= \boxed{48{,}000 \text{ psi}}$$

The total spring constant is

$$k = \frac{F}{x} = \frac{20{,}000}{4.448}$$
$$= 4496 \text{ lbf/in}$$

$$f = \frac{1}{2\pi}\sqrt{\frac{kg}{\text{weight}}}$$
$$= \frac{1}{2\pi}\sqrt{\frac{(4496)(386)}{20{,}000}}$$
$$= \boxed{1.48 \text{ Hz}}$$

NOISE CONTROL

Warmups

1. Add algebraically,

$$L_{\text{total}} = 10 \log_{10}\left[10^{\left(\frac{40}{10}\right)} + 10^{\left(\frac{35}{10}\right)}\right]$$

$$= \boxed{41.19 \text{ dB}}$$

2. If the background sound level is 43 dB,

$$45 = 10 \log_{10}\left[10^{\left(\frac{43}{10}\right)} + 10^{\left(\frac{L}{10}\right)}\right]$$

$$\text{antilog}\left(\frac{45}{10}\right) = 10^{4.3} + 10^{\left(\frac{L}{10}\right)}$$

$$11{,}670.2 = 10^{\left(\frac{L}{10}\right)}$$

$$\log_{10}(11{,}670.2) = \frac{L}{10}$$

$$L = \boxed{40.67 \text{ dB}}$$

3. Before being enclosed, the observed sound level is 100 dB. After being enclosed, the observed sound level is $110 - 30 = 80$.

The difference is $100 - 80 = \boxed{20 \text{ dB}}$

4. $\boxed{4 \text{ hours}}$

5. Using $Q = 1$ for an isotropic source, and assuming perfectly absorbing surfaces ($R = \infty$),

$$L_W = 92 - 10 \log\left(\frac{1}{4\pi(4)^2}\right) - 10.5$$

$$= 104.5 \text{ dB}$$

At $r = 12$ ft,

$$L_p = 104.5 + 10 \log\left(\frac{1}{4\pi(12)^2}\right) + 10.5$$

$$= \boxed{82.4 \text{ dB}} \qquad [\text{re } 20\,\mu\text{N/m}^2]$$

6. The general sound pressure level is

$$L_{\text{total}} = 10 \log_{10}\left[10^{\left(\frac{85}{10}\right)} + 10^{\left(\frac{90}{10}\right)}\right.$$

$$+ 10^{\left(\frac{92}{10}\right)} + 10^{\left(\frac{87}{10}\right)}$$

$$+ 10^{\left(\frac{82}{10}\right)} + 10^{\left(\frac{78}{10}\right)}$$

$$\left. + 10^{\left(\frac{65}{10}\right)} + 10^{\left(\frac{54}{10}\right)}\right]$$

$$= 95.6 \text{ dB}$$

If the A-weighted sound level is wanted (the problem is not specific), then corrections must be added to the measurements. Disregarding the 31.5 Hz frequency,

$$L_{\text{total}} = 10 \log_{10}\left[10^{\left(\frac{85-26.2}{10}\right)} + 10^{\left(\frac{90-16.1}{10}\right)}\right.$$

$$+ 10^{\left(\frac{92-8.6}{10}\right)} + 10^{\left(\frac{87-3.2}{10}\right)}$$

$$+ 10^{\left(\frac{82-0}{10}\right)} + 10^{\left(\frac{78+1.2}{10}\right)}$$

$$\left. + 10^{\left(\frac{65+1}{10}\right)} + 10^{\left(\frac{54-1.1}{10}\right)}\right]$$

$$= \boxed{88.6 \text{ dBA}}$$

7.
$$\begin{aligned} L_{500} &= 87 \\ L_{1000} &= 82 \\ L_{2000} &= \underline{78} \\ & \quad\, 247 \end{aligned}$$

$$\frac{247}{3} = \boxed{82.3}$$

8. $S/A = 0.5$ means 50% of the sound energy is removed.

$$L = 10 \log\left(\frac{0.5}{1}\right) = \boxed{-3.01} \quad [\text{decrease}]$$

Concentrates

1. $\quad A_w = \text{wall area} = (20)(100 + 100 + 400 + 400)$

$\qquad = 20{,}000 \text{ ft}^2$

$\quad A_{cf} = \text{ceiling} + \text{floor area}$

$\qquad = (2)[(100)(400)] = 80{,}000 \text{ ft}^2$

Choose NRC $= \overline{\alpha} = 0.02$ for poured concrete.

$$S_1 = (20{,}000 + 80{,}000)(0.02) = 2000$$

After treatment,

$$S_2 = (0.4)(0.8)(20{,}000) + (0.6)(0.02)(20{,}000)$$
$$+ (0.02)(80{,}000)$$
$$= 8240$$

$$\Delta L = 10\log\left(\frac{8240}{2000}\right) = \boxed{6.15 \text{ dB}}$$

2.

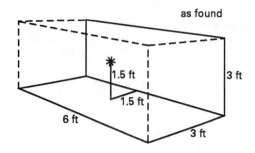

as found

area name	A	B	AB
front	18	1	18
top	18	$\frac{1}{3}$	6
sides	$\frac{18}{54}$	$\frac{1}{3}$	$\frac{6}{30}$

The existing sabin ratio is

$$R_{S_1} = \frac{30}{54} = 0.556$$

After being enclosed,

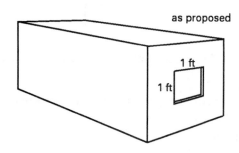

as proposed

area name	A	B	AB
side openings	2	$\frac{1}{3}$	$\frac{2}{3}$

The new sabin ratio is

$$R_{S_2} = \frac{\frac{2}{3}}{2} = 0.333$$

$$\Delta L = 10\log\left(\frac{0.556}{0.333}\right) = \boxed{2.23 \text{ dB}}$$

3.

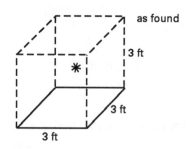

as found

area name	A	B	AB
front	9	1	9
back	9	$\frac{1}{6}$	1.5
top	9	$\frac{1}{3}$	3
sides	$\frac{18}{45}$	$\frac{1}{3}$	$\frac{6}{19.5}$

$$R_{S_1} = \frac{19.5}{45} = 0.433$$

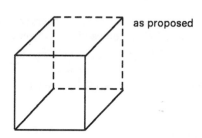

as proposed

area name	A	B	AB
back	9	$\frac{1}{6}$	1.5

$$R_{S_2} = \frac{1.5}{9} = 0.167$$

$$\Delta L = 10\log\left(\frac{0.433}{0.167}\right) = \boxed{4.14 \text{ dB}}$$

4. The sabins for each absorbing surface are:

For the floor,

$$S_1 = (1000)(0.03) = 30$$

For the ceiling (assumed smooth),

$$\alpha = 0.03$$
$$S_2 = (1000)(0.03) = 30$$

For walls and glass (both have the same $\alpha = 0.03$),

$$A = (10)(20 + 20 + 50 + 50) = 1400$$
$$S_3 = (1400)(0.03) = 42$$

For occupants, approximately 5 sabins each,

$$S_4 = (15)(5) = 75$$

For desks, approximately $1\frac{1}{2}$ sabin each,

$$S_5 = (15)\left(1\tfrac{1}{2}\right) = 22.5$$

For miscellaneous,

$$S_6 = 5 \quad \text{[given]}$$
$$S_{\text{total}} = 30 + 30 + 42 + 75 + 22.5 + 5$$
$$= 204.5$$

After treatment,

$$S_2 = (0.7)(1000) = 700$$
$$S_{\text{total}} = 30 + 700 + 42 + 75 + 22.5 + 5$$
$$= 874.5$$
$$\Delta L = 10 \log \left(\frac{874.5}{204.5}\right) = \boxed{6.31 \text{ dB}}$$

5. Originally,

$$A_{\text{ceiling}} = (20)(30) = 600$$
$$S_1 = (0.5)(600) = 300$$

After partitioning,

$$S_x = \left(\tfrac{1}{2}\right)(300) = 150$$
$$S_z = 150$$
$$A_0 = (20)(2) = 40$$

The power ratio is

$$\text{PR} = \left(\frac{\dfrac{40}{150} + \dfrac{150}{150}}{1 + \dfrac{150}{150}}\right)\left(\frac{150}{40}\right)$$
$$= 2.375$$

The insertion loss is

$$L_{\text{IL}} = 10 \log(2.375) = \boxed{3.76 \text{ dB}}$$

6.

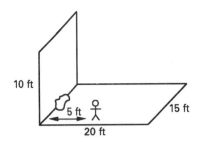

$$A_{\text{walls}} = (10)(15 + 15 + 20 + 20) = 700$$
$$A_{\text{ceiling}} = (20)(15) = 300$$
$$A_{\text{floor}} = 300$$
$$\overline{\alpha} = \frac{(700)0.06 + (300)(0.03) + (300)(0.5)}{700 + 300 + 300}$$
$$= 0.155$$
$$R = \frac{(0.155)(700 + 300 + 300)}{1 - 0.155}$$
$$= 238.5$$
$$L_p = L_w + 10 \log_{10}\left(\frac{Q}{4\pi r^2} + \frac{4}{R}\right) + 10.5$$
$$Q = 4 \quad \begin{bmatrix}\text{because the source is at the}\\ \text{intersection of 2 walls}\end{bmatrix}$$
$$L_p = 65 + 10 \log \left[\frac{4}{4\pi(5)^2} + \frac{4}{238.5}\right] + 10.5$$
$$= 60.20 \text{ dB}$$

Adding in the background noise,

$$L_{\text{total}} = 10 \log \left[10^{\left(\frac{60.20}{10}\right)} + 10^{\left(\frac{50}{10}\right)}\right]$$
$$= \boxed{60.60 \text{ dB}}$$

7. From the fan rotation,

$$\frac{600 \text{ rpm}}{60} = 10 \text{ Hz}$$

From the driving blades,

$$\frac{(600)(8)}{60} = 80 \text{ Hz}$$

From the fan blades,

$$\frac{(600)(64)}{60} = 640 \text{ Hz}$$

From the motor rotation,

$$\frac{1725}{60} \approx 29 \text{ Hz}$$

From the poles,

$$\frac{(1725)(4)}{60} = 115 \text{ Hz}$$

The electrical hum is 60 Hz.

The pulleys are

$$\text{motor pulley (same as motor)} = 29 \text{ Hz}$$
$$\text{fan pulley (same as fan)} = 10 \text{ Hz}$$

For the belt,

$$\text{belt speed} = \pi D(\text{rps}) = \pi(4) \left(\frac{1725}{60} \right)$$

$$= 361.3 \text{ in/sec}$$

The frequency is

$$\frac{361.3 \dfrac{\text{in}}{\text{sec}}}{72 \text{ in}} \approx 5 \text{ Hz}$$

8. $$f_{\text{forced}} = \frac{1725}{60} = 28.75 \text{ Hz}$$

$$f_{\text{natural}} = \frac{1}{2\pi} \sqrt{\frac{g}{\delta}} = \frac{1}{2\pi} \sqrt{\frac{386 \dfrac{\text{in}}{\text{sec}^2}}{0.02 \text{ in}}}$$

$$= 22.11 \text{ Hz}$$

The transmissibility is

$$\text{TR} = \frac{1}{\left(\dfrac{28.75}{22.11} \right)^2 - 1} = \boxed{1.448}$$

This is a 44.8% increase in force.

NUCLEAR ENGINEERING

Warmups

1. Using carbon-based amu values, the mass increase is

$$1.007825 + 1.008665 - 2.01410 = 0.00239 \text{ amu}$$

$$(0.00239 \text{ amu})\left(931.481 \frac{\text{MeV}}{\text{amu}}\right) = 2.226 \text{ MeV} \text{ [increase]}$$

Thus, $2.75 - 2.226 = 0.524$ MeV is shared by the neutron and hydrogen atom. Since both have the same mass, the neutron will receive half.

$$\boxed{0.262 \text{ MeV}}$$

2.
$$\lambda = \frac{0.693}{6.47} = 0.1071 \text{ 1/day}$$

$$0.05 = \exp\left(-0.1071t\right)$$

$$t = \boxed{27.97 \text{ days}}$$

3. Assume the shield will be 10 cm thick. For a 2 MeV source,

$$\frac{\mu_l}{\rho} = 0.0457$$

$$\mu_l = (0.0457)(11.34) = 0.5182$$

Then $\mu_l x = 5.182$ and $B = 2.78$ by interpolation. Then,

$$0.01 = 2.78 \exp\left[-(0.5182)x\right]$$

$$x = \boxed{10.86 \text{ cm}}$$

Since the initial estimate was close, a second iteration is not needed.

4.
$$\phi_u = e^{-\mu_l x} \times 10^6 = e^{-5.182} \times 10^6$$
$$= 5.62 \times 10^3 \ \gamma/\text{cm}^2\text{-s} \quad \text{[uncollided flux]}$$

The build-up flux is

$$\phi_B = 2.78\phi_u = 1.56 \times 10^4 \ \gamma/\text{cm}^2\text{-s}$$

The dose is given by $0.0659 E_o \phi_B \left(\mu_m/\rho\right)_{\text{air}}$. Since (μ_m/ρ) for air and 2 MeV gammas $= 0.238$,

$$\text{dose} = (0.0659)(2)(1.56 \times 10^4)(0.0238)$$

$$= \boxed{48.9 \text{ mR/hr}}$$

5. For 20°C gold, $\sigma_a = 98 \, b$ and $\rho = 19.32$ g/cm^3.

$$\overline{\sigma}_a = \frac{98}{1.128}\sqrt{\frac{293}{273 + 50}} = 82.75 \, b$$

$$N = \frac{(0.4909)(19.32)(6.022 \times 10^{23})}{197}$$

$$= 2.9 \times 10^{22}$$

The activated gold, Au^{198}, has a half-life of $t_{\frac{1}{2}} = 2.7d$, so

$$\lambda = \frac{0.693}{2.7} = 0.2567 \, 1/d$$

$$A = \frac{(10^8)(82.75)(10^{-24})(2.9 \times 10^{22})\left[1 - e^{(-0.2567)(1)}\right]}{3.7 \times 10^{10}}$$

$$= \boxed{1.468 \times 10^{-3} \text{ curie}}$$

6. For an isotropic point source,

$$\phi = \frac{S_o e^{-r/L}}{4\pi \overline{D} r}$$
$$\overline{D} = L^2 \Sigma_a$$
$$= (2.85)^2 \left[\frac{(0.66 \times 10^{-24})(1)(6.022 \times 10^{23})}{18}\right]$$
$$= 0.18 \quad \text{[Actually, } \overline{D} = 0.16\text{]}$$
$$\phi = \frac{(10^7)e^{-20/2.85}}{4\pi(0.16)(20)}$$

$$= \boxed{223 \text{ neutrons/cm}^2\text{-s}}$$

7.
$$\sigma_f = 4.18 \, b$$
$$\sigma_a = 7.68 \, b$$

$$p(\text{fission}) = \frac{4.18}{7.68}$$

$$= \boxed{0.544 \ (54.4\%)}$$

8. Assume the shield is 20 cm thick. For 1 MeV gamma in iron, $\mu_l/\rho = 0.0595$. For iron, $\rho = 7.87$, so $\mu_l = 0.4683$ and $\mu_l x = 9.366$. $B = 14.93$ by interpolation.

$$\phi_B = \frac{(14.93)(10^8)e^{-0.4683x}}{4\pi x^2}$$

$$= \frac{1.19 \times 10^8 e^{-0.4683x}}{x^2}$$

The exposure rate is

$$1 = \frac{(0.0659)(1)(1.19 \times 10^8)e^{-0.4683x}(0.0280)}{x^2}$$

By trial and error,

$$x = \boxed{14.8 \text{ cm}}$$

Concentrates

1. If the conversion ratio is not assumed, calculate for fast fission.

$$U\text{-}238: \sigma_a = 0.59\,b; \quad \sigma_f = 0.5\,b$$

$$P_U\text{-}239: \sigma_a = 1.95\,b; \quad \sigma_f = 1.8\,b$$

$$\eta = (0.8)(2.45)\left(\frac{0.5}{0.59}\right) + (0.2)(2.95)\left(\frac{1.8}{1.95}\right)$$

$$= 2.206$$

Assume $\epsilon = 1.05$, so the conversion ratio is

$$CR = (2.206)(1.05) - 1 = 1.316$$

$$\text{linear } t_d = \frac{(2000)(1000)}{(1.316-1)(1.23)(1000)} = 5146 \text{ days}$$

$$\text{exponential} = (0.693)(5146) = \boxed{3566 \text{ days}}$$

2. $$P = \frac{\overline{\phi}\Sigma_f V}{3.1 \times 10^{10}}$$

$$= \frac{\left(\frac{4.5 \times 10^{15}}{3.29}\right)(0.005)\left[\frac{4}{3}\pi(40)^3\right]}{3.1 \times 10^{10}}$$

$$= \boxed{5.91 \times 10^7 \text{ W}}$$

3. $$r_1 = \sqrt{\frac{(25.4)^2}{\pi}} = 14.33 \text{ cm}$$

$$E = 1 + \frac{(14.33)^2}{(2)(59)^2}$$

$$\times \left[\frac{\ln\left(\frac{14.33}{1.02}\right)}{1 - \left(\frac{1.02}{14.33}\right)^2} - 0.75 + \left(\frac{1.02}{(2)(14.33)}\right)^2 \right]$$

$$= 1.0563$$

$$F \approx 1 + \left(\tfrac{1}{2}\right)\left(\frac{r_o}{2L}\right)^2 - \left(\frac{1}{12}\right)\left(\frac{r_o}{2L}\right)^4$$

$$+ \left(\frac{1}{48}\right)\left(\frac{r_o}{2L}\right)^6$$

$$L = 1.55 \text{ cm} \quad \text{[for natural radiation]}$$

$$F \approx 1 + \left(\tfrac{1}{2}\right)\left[\frac{1.02}{(2)(1.55)}\right]^2 - \left(\frac{1}{12}\right)\left[\frac{1.02}{(2)(1.55)}\right]^4$$

$$+ \left(\frac{1}{48}\right)\left[\frac{1.02}{(2)(1.55)}\right]^6$$

$$= 1.0532$$

$$\Sigma_{a,f} = \frac{(18.7)(6.022 \times 10^{23})(7.68 \times 10^{-24})}{(237.98)(1.128)} \times (0.984)$$

$$= 0.3170$$

$$\Sigma_{a,m} = \frac{(1.6)(6.022 \times 10^{23})(4.65 \times 10^{-27})}{(12)(1.128)}$$

$$= 3.310 \times 10^{-4}$$

$$V_f \propto \pi r_o^2 = \pi(1.02)^2 = 3.2685$$

$$V_m = (25.4)^2 - 3.2685 = 641.8915$$

$$\frac{\overline{\phi}_m}{\overline{\phi}_f} = 1.0532 + \left(\frac{3.2685}{641.8915}\right)\left(\frac{0.3171}{3.311 \times 10^{-4}}\right)$$

$$\times (0.0563)$$

$$= 1.328$$

$$\frac{1}{f} = 1 + (1.328)\left(\frac{3.311 \times 10^{-4}}{0.3171}\right)\left(\frac{641.8915}{3.2685}\right)$$

$$= 1.2723$$

$$f = \boxed{0.7860}$$

Timed

1. The mass of uranium is

$$m = (100{,}000 \text{ lbm})\left(0.4536 \frac{\text{kg}}{\text{lbm}}\right) = 4.54 \times 10^4 \text{ kg}$$

The number of fuel molecules per unit volume is

$$N_f = \frac{(0.02)(4.54 \times 10^4 \text{ kg})\left(1000 \frac{\text{g}}{\text{kg}}\right) \times \left(6.022 \times 10^{23} \frac{\text{particles}}{\text{gmole}}\right)}{(V \text{ cm}^3)\left(235 \frac{\text{g}}{\text{gmole}}\right)}$$

$$= \frac{2.33 \times 10^{27}}{V} \text{ particles/cm}^3$$

From the thermal power equation,

$$\phi = \frac{(3.1 \times 10^{10})(P_{\text{th}})}{\sigma_f N_f V}$$

$$= \frac{\left(3.1 \times 10^{10}\,\dfrac{\frac{\text{fissions}}{\text{sec}}}{\text{W}}\right)(5.0 \times 10^8\ \text{W})}{\left(547\,\dfrac{\text{barns}}{\text{particle}}\right)\left(1 \times 10^{-24}\,\dfrac{\text{cm}^2}{\text{barn}}\right)}$$
$$\times \left(2.33 \times 10^{27}\text{particles}\right)$$

$$= \boxed{1.22 \times 10^{13}\ \text{fissions/cm}^2\text{-sec}}$$

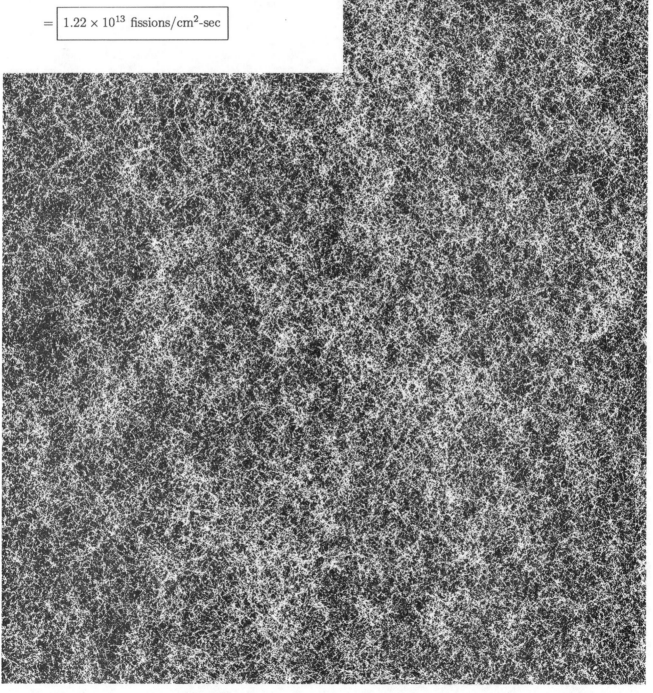

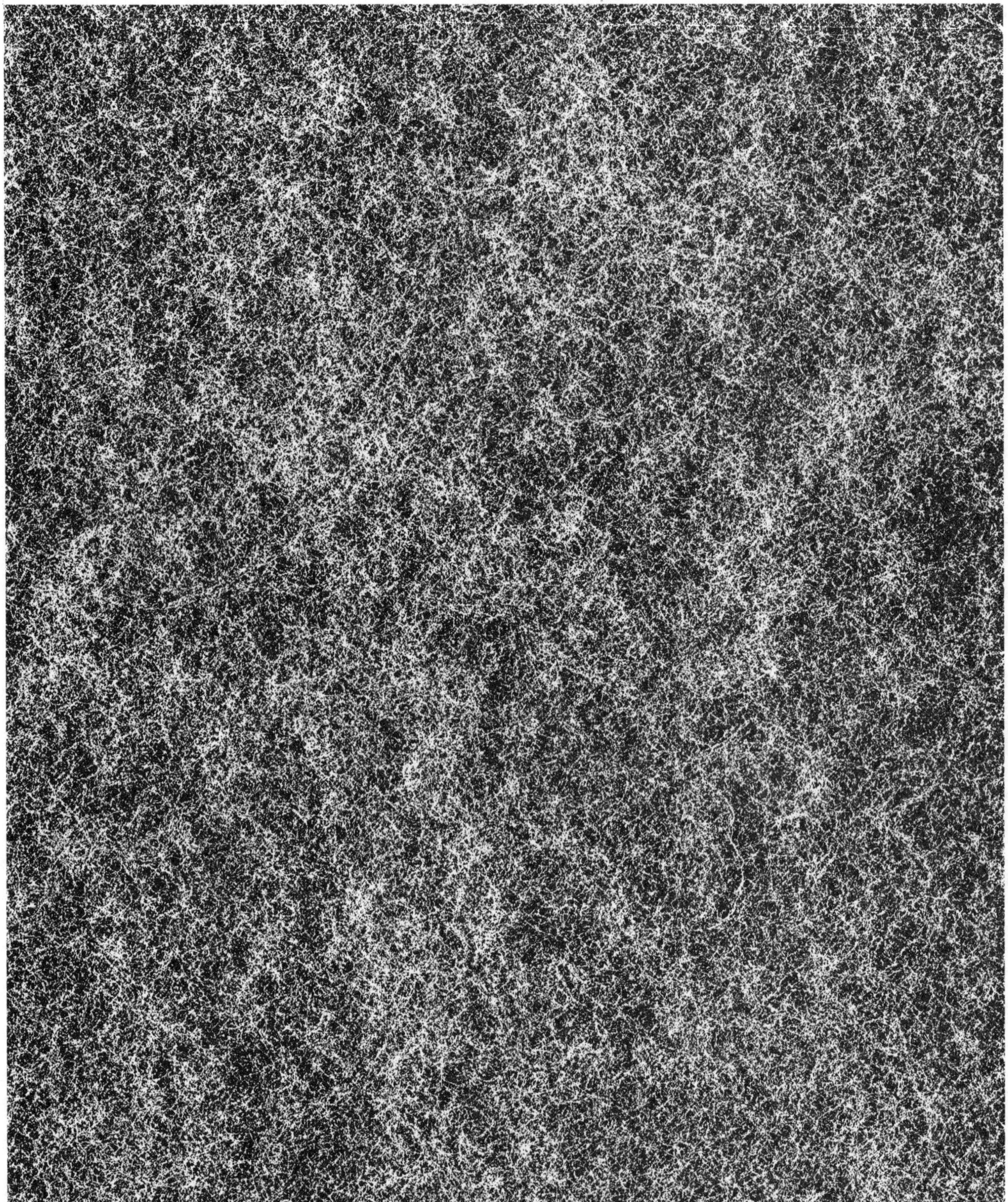

MODELING OF ENGINEERING SYSTEMS

Warmups

1.

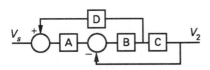

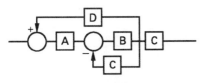

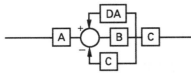

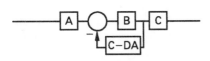

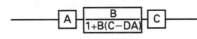

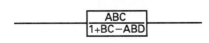

2.

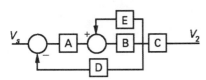

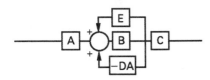

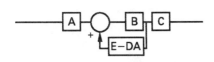

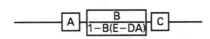

3.1.

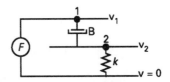

At node 1,

$$F = B(v_1 - v_2) \quad \text{or} \quad F = B(x_1' - x_2')$$

At node 2,

$$F = k(x_2 - 0) \quad \text{or} \quad 0 = kx_2 + B(x_2' - x_1')$$

3.2.

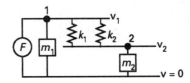

At node 1,

$$F = m_1 a_1 + (k_1 + k_2)(x_1 - x_2)$$
$$= m_1 x_1'' + (k_1 + k_2)(x_1 + x_2)$$

At node 2,

$$0 = m_2 a_2 + (k_1 + k_2)(x_2 - x_1)$$
$$= m_2 x_2'' + (k_1 + k_2)(x_2 - x_1)$$

3.3. Despite its appearance, this is a rotational system.

$$\tau = \text{applied rotational torque} = FL$$
$$\theta = \text{rotated angle}$$
$$x_1 = \text{arc distance}$$
$$x_2 = \text{arc distance}$$
$$I = \text{moment of inertia of beam about an end}$$
$$= \tfrac{1}{3}mL^2$$

PROFESSIONAL PUBLICATIONS, INC. ● Belmont, CA

$m =$ resisting moment $= kx_2l$

$\quad = kl \sin \theta l$

$\quad = kl^2\theta \qquad$ [for small values of θ]

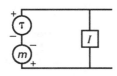

$$\tau = I\alpha$$

$$FL - kl^2\theta = \tfrac{1}{3}mL^2\alpha$$

$$\alpha = \theta''$$

$$\theta'' + \left(\frac{3kl^2}{mL^2}\right)\theta = \frac{FL}{I}$$

3.4.

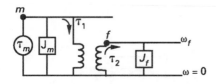

At node m,

$$\tau_m = J_m\alpha_m + \tau_1$$

$$\quad = J_m\theta''_m + \tau_1$$

At node f,

$$\tau_2 = J_f\alpha_f$$

$$\tau_2 = \left(\frac{n_2}{n_1}\right)\tau_1$$

$$\left(\frac{n_2}{n_1}\right)\tau_1 = J_f\theta''_f$$

The third equation required is

$$\theta_m = \left(\frac{n_2}{n_1}\right)\theta_f$$

3.5. This is an example of an untuned vibration damper. Consider the fluid to act as a damper with coefficient B.

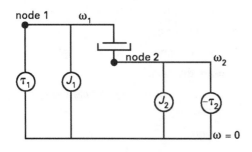

For node 1,

$$\tau_1 = J_1\alpha_1 + B(\omega_1 - \omega_2) = J_1\theta''_1 + B(\theta'_1 - \theta'_2)$$

For node 2,

$$-\tau_2 = J_2\alpha_2 + B(\omega_2 - \omega_1) = J_2\theta''_2 + B(\theta'_2 - \theta'_1)$$

$\tau = B(\omega_1 - \omega_2)$, so the third equation is

$$B = \frac{\tau}{\omega_2 - \omega_1} = \frac{\tau}{\theta'_2 - \theta'_1}$$

3.6.

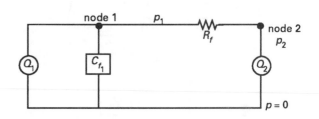

$$Q = A_t\left(\frac{dh}{dt}\right) = \left(\frac{A_t}{\rho}\right)\left(\frac{dp}{dt}\right)$$

$$\quad = C_f\left(\frac{dp}{dt}\right)$$

For node 1,

$$Q_1 = C_{f_1}\left(\frac{dp_1}{dt}\right) + \left(\frac{1}{R_f}\right)(p_2 - p_1)$$

For node 2,

$$Q_2 = \left(\frac{1}{R_f}\right)(p_1 - p_2)$$

3.7. A pump is required to fill the tank against the static head.

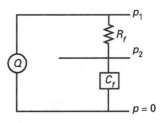

The system equations are

$$Q = \left(\frac{1}{R_f}\right)(p_1 - p_2) = C_f\left(\frac{dp_2}{dt}\right)$$

Timed

1. First, simplify the system.

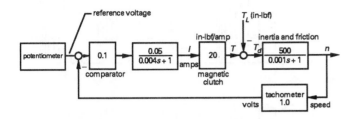

Redraw the system in more tradtional form.

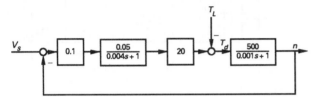

Simplify.

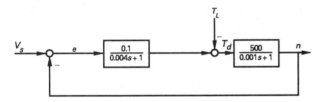

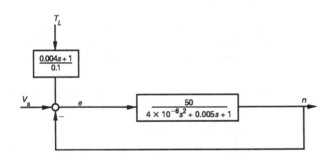

(a) Disregarding the feedback loop, there are two open-loop transfer functions: one each from inputs V_s and T_L.

$$\frac{n}{V_s} = \frac{50}{(0.001s + 1)(0.004s + 1)}$$

$$\frac{n}{T_L} = \frac{500}{0.001s + 1}$$

The total open-loop response is the sum of the responses to the two inputs.

$$n = \frac{50V_s}{(0.001s + 1)(0.004s + 1)} + \frac{500T_L}{0.001s + 1}$$

The two transfer functions have been plotted below; the open-loop steady-state gain (OLSSG) can be read off the plots or obtained by setting $s = 0$ in the open-loop transfer equations.

$$\text{OLSSG}(V_s) = 50$$

$$\text{OLSSG}(T_L) = 500$$

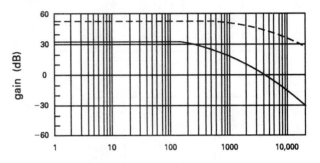

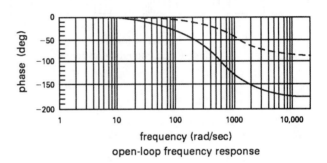

open-loop frequency response

(b) Closing the loop using negative feedback and simplifying results in

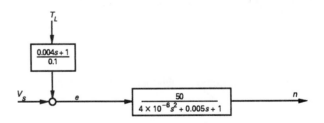

Again, there are two transfer functions: one for V_s and one for T_L. The sum of the two is the total closed-loop response.

$$n = \frac{V_s}{\dfrac{(0.001)(0.004)s^2}{50} + \dfrac{(0.001 + 0.004)s}{50} + 1}$$

$$+ \frac{(10)(0.004s + 1)T_L}{\dfrac{(0.001)(0.004)s^2}{50} + \dfrac{(0.001 + 0.004)s}{50} + 1}$$

The closed-loop responses are

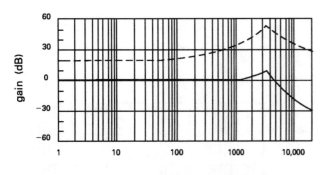

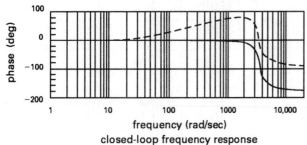

frequency (rad/sec)
closed-loop frequency response

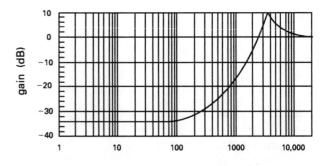

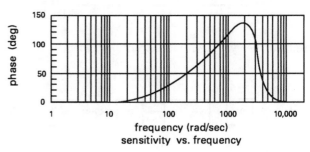

frequency (rad/sec)
sensitivity vs. frequency

The response of n to V_s is unity and flat out to 1000 rad/sec. Also, note the low frequency response of n to T_L has been reduced from 500 (54 dB) to 10 (20 dB). Both of these are desirable effects of closing the loop.

We also see that one of the effects of closing the loop is the change the steady-state responses, V_s to $n = 1.0$, and T_L to $n = 10$.

(c) This is a negative feedback system. The sensitivity is

$$S = \frac{1}{1 + GH}$$

The sensitivity for the V_s to n transfer function is

$$S = \frac{(0.004s + 1)(0.001s + 1)}{(0.004s + 1)(0.001s + 1) + 50}$$

$$= \frac{(0.02)(0.004s + 1)(0.001s + 1)}{\dfrac{s^2}{3536} + \dfrac{(2)(0.176)s}{3536} + 1}$$

(d) From the equation for the total closed-loop response, n, the response due to a step in V_s will be second order. Putting the equation for n into standard form (with $T_L = 0$),

$$n = \frac{V_s}{\dfrac{s^2}{(3536)^2} + \dfrac{(2)(0.176)s}{3536} + 1}$$

This shows that the natural frequency is $\omega = 3536$ 1/sec, and the damping factor is $\xi = 0.176$.

This means that the response to a step will be fast (90° rise time of 0.4 ms) but will have significant overshoot (about 55%). The response will be oscillatory since $\xi < 1.0$.

As long as $T_L = 0$, there will not be any steady-state error.

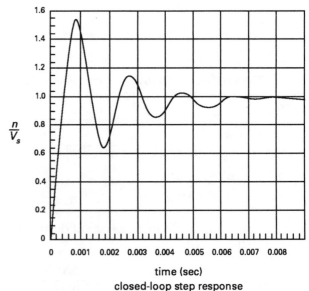

time (sec)
closed-loop step response

(e) The closed-loop response to a step load change will be similar to the step change in V_s except the numerator in the transfer function, n, will cause the response to deviate from second order.

However, the response will still be oscillatory, and there will be overshoot. Also, since $T_L \neq 0$, there will be an error from the setpoint, V_s.

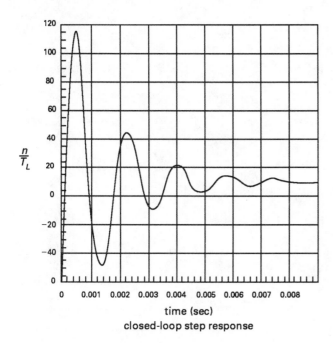

closed-loop step response

(f) Let the comparator gain equal k, and resolve for the open-loop transfer function for V_s.

$$\frac{n}{V_s} = \frac{500k}{(0.001s + 1)(0.001s + 1)}$$

Closing the loop yields

$$\frac{n}{V_s} = \frac{1}{(1 \times 10^{-6}) s^2 + 0.005s + 500k}$$

This is a simple second-order problem. There are only two roots, and the system will be stable for any $k > 0$ and will be unstable for $k < 0$.

Practically, elements of the physical system will limit the gain k, but such limitations are not part of the linear model.

(g) The system steady-state response can be improved by adding integral control. This will effectively compensate for any steady-state disturbance due to T_L. This addition, however, has a side effect of reducing the stability margin of the system. However, if properly designed, the system will still be stable.

2. This is a standard second-order system.

$$F = mx'' + Bx' + k_s x$$

The natural frequency is

$$\omega = \sqrt{\frac{k_s}{m}}$$

$$= \sqrt{\frac{1200 \frac{\text{lbf}}{\text{ft}}}{(100 \text{ lbm})\left(\frac{1}{32.2}\frac{\text{lbf-sec}^2}{\text{lbm-ft}}\right)}}$$

$$= \boxed{19.6 \text{ rad/sec (3.13 Hz)}}$$

$$\xi = \frac{B}{2\sqrt{k_s m}} = \frac{60 \frac{\text{lbf-sec}}{\text{ft}}}{2\sqrt{\left(1200 \frac{\text{lbf}}{\text{ft}}\right)\left(\frac{100 \text{ lbm}}{32.2 \frac{\text{lbm-ft}}{\text{lbf-sec}^2}}\right)}}$$

$$= \boxed{0.49}$$

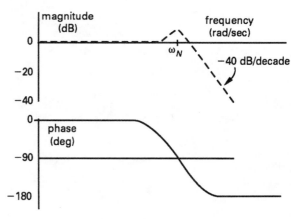

frequency response

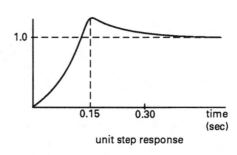

unit step response